A BRIEF HISTORY OF EVERYONE WHO EVER LIVED

Also by Adam Rutherford

Creation

A BRIEF HISTORY OF EVERYONE WHO EVER LIVED

The Human Story Retold Through Our Genes

ADAM RUTHERFORD

FOREWORD BY SIDDHARTHA MUKHERJEE

NEW YORK

A Brief History of Everyone Who Ever Lived: *The Human Story Retold Through Our Genes*

Originally published in Great Britain by Weidenfeld & Nicolson, an imprint of the Orion Publishing Group Ltd., a Hachette UK company, in 2016.
First published in North America by The Experiment, LLC, in 2017.

The Experiment, LLC
220 East 23rd Street, Suite 301
New York, NY 10010-4674
theexperimentpublishing.com

Library of Congress Cataloging-in-Publication Data

Names: Rutherford, Adam, author.
Title: A brief history of everyone who ever lived : the human story retold through our genes / Adam Rutherford.
Description: New York : The Experiment, 2017. | Includes bibliographical references and index.
Identifiers: LCCN 2017022566| ISBN 9781615194049 (hardcover) | ISBN 9781615194186 (ebook)
Subjects: | MESH: Genome, Human--genetics | Genomics--history | Biological Evolution | DNA--history
Classification: LCC QH445.2 | NLM QU 11.1 | DDC 611/.0181663--dc23
LC record available at https://lccn.loc.gov/2017022566

ISBN 978-1-61519-404-9
Ebook ISBN 978-1-61519-418-6

Cover design by Pete Garceau
Text design by Sarah Schneider

Manufactured in the United States of America
Distributed by Workman Publishing Company, Inc.
Distributed simultaneously in Canada by Thomas Allen & Son Ltd.

First printing October 2017
10 9 8 7 6 5 4 3 2

I was tutored and schooled by Professor Steve Jones, at University College London and beyond. On the first day of his undergraduate genetics course in 1994, he offered to compensate any of us impoverished students the profit if we bought a copy of his masterpiece *The Language of the Genes.* I claimed that 55 pence. Over the years he has influenced me intellectually perhaps more than anyone else, and in many ways, this book is, with his permission, a continuation of that classic. In 2012, when I was invited to give a prestigious lecture for the British Humanist Association, Steve introduced me. He joked, I hope, that he had a strong sense that I was waiting for him to die so I could truly inherit his living. Because he's still not dead, and for the 55 pence, I dedicate this book to

STEVE JONES

CONTENTS

Foreword by Siddhartha Mukherjee ix
Author's note xiii

Introduction 1

PART ONE: HOW WE CAME TO BE

1. Horny and mobile 14
2. The first European union 65
3. These American lands 128
4. When we were kings 157

PART TWO: WHO WE ARE NOW

5. The end of race 214
6. The most wondrous map ever produced by humankind 265
7. Fate 311
8. A short introduction to the future of humankind 339

Epilogue 361

Acknowledgments 364
Glossary 366
References and further reading 370
Text and image credits 385
Index 386
About the Author 402

FOREWORD
BY SIDDHARTHA MUKHERJEE

It is humbling to introduce the North American edition of Adam Rutherford's monumental *A Brief History of Everyone Who Ever Lived.* Ambitious, wide-ranging, and deeply researched, Rutherford's book sets out to describe the history of the human species—from our origins as a slight, sly, naked, apelike creature somewhere in Africa to our gradual spread across the globe and our dominion over the planet. In academic circles, it is becoming fashionable to use the word *anthropocene* to describe the current epoch, when humans have begun to have significant effect on the earth and the environment. We have, in short, reshaped the world that we live in, but Rutherford turns our attention to a somewhat different, if equally essential, question: How did the world—biology, environment, culture—shape *us*?

The key idea that Rutherford unveils in this riveting volume is that human genomics—the study of our DNA—is radically altering our understanding of our own past. Traditionally, we have investigated questions of human origin by studying biological and cultural artifacts—skeletons, tools, architectural remains, books, stories, language, rituals. But the genome, Rutherford argues, is also an artifact: It stores powerful information about heritage, enabling scientists to reconstruct human origin based on that information alone. Did the first settlers in North America arrive across the Bering Strait several thousand years ago? Where and when did *Homo sapiens* coexist and interbreed with Neanderthals?

How old are we as a species, and how, exactly, do we define where we were "born"? The study of human DNA is unveiling astonishingly novel insights into such questions, Rutherford writes. Indeed, one of the most surprising features of the *genetic* investigation of human history (as opposed to more traditional means of approaching the question) is the number of myths and fallacies that human genomics has already overturned, and how much of what seemed known and well established is, in fact, unknown and steeped in ambiguity.

In this edition, Rutherford tackles a few thorny concerns that are particularly relevant to our side of the world. Our attempt to reconstruct the early history of human settlement in North America has been dramatically reshaped by modern genomics. By studying genes, we might be able to understand the migration patterns of humans across this continent, decipher the lineal relationships between tribes, and even track the first genetic intersections between Native Americans and European settlers. But this knowledge has not been easy to come by, Rutherford argues. We know much less about American history than we *could* know because of the unique manner in which the United States developed. The profound failures in the relationships between Native Americans and European settlers—failures driven, in large part, by the toxic legacy of colonialism—have made it impossible, at times, to answer some of the fundamental questions about the origins of humans in the Americas. It is a strange shame that our cultural history has made it monumentally difficult to unearth our biological and anthropological history.

Rutherford is hopeful that this shadow of suspicion and distrust will finally ease, enabling scientists to be able to train the ever-expanding—and ever-more-acute—lenses of genome technologies on the history of the Americas. The capacity to sequence, analyze, and store vast troves of genetic information has made it possible, in principle, to answer deep questions about American history. The reconstruction of ancestry—previously a parlor game that could only be played by the ultrarich—is being popularized and commercialized: With a cheek swab, a drop of saliva, and a few hundred dollars (as the infomercials on TV and the web exhort us), we can now easily obtain information about our individual heritage. But the reconstruction of our national past is unlikely to be easy, he writes. To make inroads into this uncharted continent of genomics, we must first tackle the legacy of European colonialism with caution, openness, and fortitude. This kind of wisdom—rarely to be found in academic textbooks of genetics—catapults Rutherford's book beyond the realm of popular science writing into the domains of philosophy, history of science, and cultural studies.

The study of DNA—the molecule that stores information about heritage—is a rather modern idea. (Indeed, we did not know that DNA was the carrier of genetic information until the 1950s. The elegant double-helix structure of the most beautiful molecule in biology was solved just sixty-four years ago, and the genetic code was only deciphered in the early '60s.) But the desire to understand heritage, Rutherford reminds us, is an ancient desire—and twisted into that desire are our concerns about identity and relationships, and our sense of self. As Rutherford concludes, we cannot investigate "heritage" simply by studying DNA; we also need to understand the social and political history of heritage. In this endlessly intriguing book, we are thus not just presented with the mini-history of the human genome, but also with a sweeping history of our attempt to grapple with the human genome.

SM

June 2017

Siddhartha Mukherjee is the author of *The Gene* and the Pulitzer Prize–winning *The Emperor of All Maladies.*

AUTHOR'S NOTE

Science demands collaboration. There are no lone geniuses, never evil geniuses, and very rarely any heretical geniuses. Almost all science is done by very normal people working in teams or in cahoots with others in similar or dissimilar fields, and they build knowledge on the shoulders of historical and contemporary giants, as Isaac Newton once suggested, parroting the words of the eleventh-century philosopher Bernard of Chartres, who was referencing the Greek myth of the temporarily blinded hunter Orion, who saw further by sitting a dwarf on his shoulders.

The science in this book is perhaps more collaborative than most, as it involves the introduction of a new discipline, genomics, into older ones, namely history, archaeology, paleoanthropology, medicine, and psychology. Author lists of genetics papers can now run into the dozens, hundreds, and occasionally thousands. Long gone are the days when Victorian gentlemen could idle away their inheritances in hot pursuit of the fabric of nature.

Many people have helped me with the writing of this book, and I have used numerous research papers, which are listed at the back. For the most part, though, I have not included specific references in the text, nor individual researchers, simply to add to the flow of the stories herein. A large number of the studies involve Mark Thomas at University College London, and I am very grateful for his guidance and friendship over the years. The particular field of ancient

DNA is led by a few labs currently, though it is spreading at a feverish pace as the techniques become better and easier to deploy, and as more and more data is accrued. Several of these tales are drawn from the work of Svante Pääbo, Turi King and the Richard III project, Joe Pickrell, David Reich, Josh Akey, Joachim Burger, Graham Coop, Johannes Krause, and a few others, who have all helped me directly or indirectly. The work is theirs; any errors are mine. On page 366, there is a glossary of some of the technical or less than friendly terms that geneticists use.

Introduction

> "In the distant future I see open fields for far more important researches . . . Light will be thrown on the origin of man and his history."
>
> "Chapter 14: Recapitulation and Conclusion" in *On the Origin of Species* by Charles Darwin, 1859

This is a story about you. It concerns the tale of who you are and how you came to be. It is your individual story, because the journey of life that alights at your existence is unique, as it is for every person who has ever drawn breath. And it's also our collective story, because as an ambassador for the whole of our species, you are both typical and exceptional. Despite our differences, all humans are remarkably close relatives, and our family tree is pollarded, and tortuous, and not in the slightest bit like a tree. But we are the fruit thereof.

Something on the order of 107 billion modern humans have existed, though this number depends on when exactly you start counting. All of them—of us—are close cousins, because our species has a single African origin. We don't quite have the language to describe what that really means. It doesn't, for example, mean a single couple, a hypothetical Adam and Eve. We think of families and pedigrees and genealogies and ancestry, and we try to think of the deep past in the same way. Who were my ancestors? You might have a simple, traditional family structure or, one like mine, handsomely untidy, its tendrils jumbled like old wires in a drawer. But no matter which, everyone's past becomes muddled sooner or later.

We all have two parents, and they had two parents, and all of them had two parents, and so on. Keep going like this all the way back to the last time England was invaded, and you'll see that doubling each generation results in more people than have ever lived, by many billions. The truth is that our pedigrees fold in on themselves, the branches loop back and become nets, and all of us who have ever lived have done so enmeshed in a web of ancestry. We only have to go back a few dozen centuries to see that most of the 7 billion of us alive today are descended from a tiny handful of people, the population of a village.

History is the stuff that we have recorded. For thousands of years, we have painted, carved, written, and spoken the stories of our pasts and presents, in attempts to understand who we are and how we came to be. By consensus, history begins with writing. Before that we have prehistory—the stuff that happened before we wrote it down. For the sake of perspective, life has existed on Earth for about 3.9 billion years. The species *Homo sapiens*, of which you are a member, emerged a mere 300,000 years ago, as far as we know, in pockets in the east and north of Africa. Writing began about 6,000 years ago, in Mesopotamia, somewhere in what we now call the Middle East.

For comparison, the book you are holding is around 115,000 words, or 685,000 characters long, including spaces. If the length of time life has existed on Earth were represented as this book, each character, including spaces, is around 5,957 years. Anatomically modern humans' tenure on Earth is equivalent to

. . . the precise length of this phrase.

The time we have been recording history is an evolutionary wing-flap equivalent to a single character, the width of this period<.>

And how sparse that history is! Documents vanish, dissolve, decompose. They are washed away by the weather, or consumed by insects and bacteria, or destroyed, hidden, obfuscated, or revised.

That is before we address the subjectivity of the historical record. We can't agree definitively on what happened in the last decade. Newspapers record stories with biases firmly in place. Cameras record images curated by people and only see what passes through the lens, frequently without context. Humans themselves are terribly unreliable witnesses to objective reality. We fumble.

The precise details of the events of September 11, 2001, when the World Trade Center towers were destroyed, may well remain obscure because of conflicting reports and the chaos of those horrors. Witness testimonies in courts are notoriously defective and are always subject to squint-eye scrutiny. Flit back a few centuries, and there is no contemporary evidence even for the existence of Jesus Christ, arguably the most influential man in history. Most of our tales about his life were written in the decades after his death by people who had never met him. Today, we would seriously question that, if it were presented as historical evidence. Even the accounts that Christians rely on, the Gospels, are inconsistent and have irreversibly mutated over time.

This is not to disparage the study of history (nor Christianity). It's merely a comment on how the past is foggy. Until recently it was recorded primarily in religious texts, business transaction documents, and the papers of royal lineages. In modern times we have the opposite problem—far too much information and almost no way to curate it. In every purchase you make online, every Internet search you do, you volunteer information about yourself to be captured by companies in the ether. Books, sagas, oral histories, inscriptions, archaeology, the Internet, databases, film, radio, hard drives, tape. We piece together these bits and bytes of information to reconstruct the past. And now, biology has become part of that formidable swill of information.

The epigraph at the beginning of this introduction is Darwin's single reference to humans in *On the Origin of Species*, right at the end, as if to tease us that there will be a sequel. With his proposed theory of descent with modification in the distant future, light will be shed on our own story: to be continued.

That time has come. There is now another way to read our pasts, and floodlights are being shone on our origins. You carry an epic poem in your cells. It's an incomparable, sprawling, unique, meandering saga. About a decade ago, fifty years after the discovery of the double helix, our ability to read DNA had improved to the degree that it was transformed into a historical source, a text to pore over. Our genomes, genes, and DNA house a record of the journey that life on Earth has taken—4 billion years of error and trial that resulted in you. Your genome is the totality of your DNA, 3 billion letters of it, and due to the way it comes together—by the mysterious (from a biological point of view) business of sex—it is unique to you. Not only is this genetic fingerprint yours alone, it's unlike any other of the 107 billion people who have ever lived. That applies even if you are an identical twin, whose genomes begin their existence indistinguishable, but inch away from each other moments after conception. In the words of Dr. Seuss:

> Today you are you! That is truer than true!
> There is no one alive who is you-er than you!

The sperm that made you started its life in your father's testicles within a few days before your conception. One single sperm out of a spurt of billions ground its head against your mother's egg, one of just a few hundred. Like a Russian doll, that egg had grown in her when she was growing inside her mother, but it matured within the last menstrual cycle and, taking its turn from alternating ovaries, eased its way out of the comfort of its birthplace. On contact, that winning sperm released a chemical that dissolved the egg's reluctant membrane, left its whiplash tail behind, and burrowed in.

Once inside, the egg set an impenetrable fence that stopped any others breaching her defenses. The sperm was unique, as was the egg, and the combination of the two, well, that was unique too, and that became you. Even the point of entry was unique. Your mother's egg being roughly spherical, that sperm could've punched its way in anywhere, and at the behest of cosmic happenstance, it penetrated its quarry at a singular point, a point that set waves of chemicals and effectively began the process of setting your body plan—head at one end, tail at the other. In other organisms, we know that if the winning sperm had come in on the other side, the embryo that became you would've started growing in a different orientation, and it may well be the same in us.

Your parents' genetic material, their genome, had been shuffled in the formation of sperm and egg, and halved. Their parents, your grandparents, had provided them with two sets of chromosomes, and the shuffle mixed them up to produce a deck that had never existed before, and never will again. They also bestowed upon you just a bit of unshuffled DNA. If you're a man, you have a Y chromosome that was largely unchanged from your father and from his father and so on back through time. It's a stunted shriveled piece of DNA, with only a few genes on it and a lot of debris. The egg also had some small loops of DNA hiding inside, in its mitochondria, tiny powerhouses that provide power for all cells. It has its own mini genome, and because it sits inside the egg, this only comes from mothers. Together, these two make up a tiny proportion of your total DNA, but their clear lineages have some use when tracking back through genealogies and ancient history. However, the vast majority of your DNA was forged in the shuffle of your parents', and theirs in theirs. That process happened every time a human lived; the chain that precedes you is unbroken.

> They fuck you up, your mum and dad.
> They may not mean to, but they do.
> They fill you with the faults they had
> And add some extra, just for you.

I offer no comment on the psychological or parental aspects of Philip Larkin's poem, but from a biological point of view, it's spot on. Each time an egg or sperm is made, the shuffle produces new variation, unique differences in the people that host them. You'll inherit your parents' DNA in unique combinations, and in that process—meiosis—you also will have invented some brand new genetic variations, just for you. Some of those will get passed on if you have children, and they will acquire their own as well.

It's upon these differences in populations that evolution can act, and it's in these differences that we can follow the path of humankind, as we have roamed across land and oceans, and oceans of time, into every corner of the planet. Geneticists have suddenly become historians.

A single genome contains a huge amount of uncurated data, enough to lay out plans for a human. But genomics is a comparative science. Two sets of DNA from different people contain much more than double that information. All human genomes host the same genes, but they all may be slightly different, which accounts for the fact that we are all incredibly similar, and utterly unique. By comparing those differences we can make inferences about how closely related those two people are, and when those differences evolved. We can now extend these comparisons to all humanity, as long as we can pull DNA from your cells.

When the first complete human genome was published in 2001 to great fanfare, it was in fact a sketchy draft readout of most of the genetic material of just a few of us. To get this far had taken hundreds of scientists the best part of a decade, and had cost on the order of $3 billion, approximately $1 per letter of DNA. Just fifteen years later, things are emphatically easier, and the amount of data from individual genomes now is incalculable. As I write these words we have approximately 150,000 fully sequenced human genomes, and useful samplings from literally millions of people, from all over the world. Grand medical endeavors with accurate names like "The Hundred Thousand Genome Project" typify how easily we can now extract the data that we all store in our living cells. Here in the UK, we are seriously considering sequencing genomes of everyone at birth. And it's not limited to the rigor of formal science or governmental medical policy: You can spit in a test tube and get a read-out of key parts of your own genome from an armada of companies that will tell you all sorts of things about your characteristics, history, and risk of some diseases, for just a couple of hundred dollars.

We now have genomes of hundreds of long dead people too to slot into this grand narrative. The bones of an English king, Richard III, were identified in 2014 with a raft of archaeological evidence (Chapter 4), but the deal was royally sealed with his DNA. The kings and queens of the past are known to us because of their status, and because history is dominated by telling and retelling their stories. While genetics has enriched the study of monarchs, DNA is the ultimate leveler, and our newfound ability to extract the finest details of the living past has rendered this an examination of the people, of countries, of migration, of everyone. We can test, and verify or falsify, and know the histories of the *people*, not just the powerful or the celebrities of their day. Nobodies from the past are being elevated to some of the most important people who ever lived. DNA is universal and, as we'll find out, being in a royal lineage might afford you divine rights over citizens, and the spoils that go with inherited power, but evolution, genetics, and sex are largely indifferent to nationalities, borders, and all that heady power.

And we can look further still. The study of ancient humans was once limited to old teeth and bones and the ghostly traces of their lives left in dirt, but we can now piece together the genetic information of truly ancient humans, of Neanderthals and other extinct members of our extended family, and these people are revealing a new route to where we are today. We can pluck out their DNA to tell us things that could not be known in any other way—we can, for example, know how a Neanderthal person experienced smell.

Retrieved after epochs, DNA has profoundly revised our evolutionary story. The past may be a foreign country, but the maps were inside us the whole time.

The amount of data this new science is generating is colossal, phenomenal, overwhelming. Studies are being published every week that upend what has come before. In the penultimate stages of writing this book, the date of the great exodus from Africa may have shifted more than 10,000 years earlier than previously thought, following the discovery of forty-seven modern teeth in China. Then in the final stages it moved back by another 20,000 years with the detection of *Homo sapiens* DNA in a millennia-dead Neanderthal girl. These numbers are not much in evolutionary terms, ripples in geological time. But that is much more than the whole of written human history, and so the land continually and dramatically moves under our feet.

The first half of this book is about the rewriting of the past using genetics, from a time when there were at least four human species on Earth right up to the kings of Europe into the eighteenth century. The second half is about who we are today, and what the study of DNA in the twenty-first century says about families, health, psychology, race, and the fate of us. Both parts are drawn from using DNA as a text to sit alongside the historical sources we have relied on for centuries: archaeology, rocks, old bones, legends, chronicles, and family histories.

Although the study of ancestors and inheritance is as old as humans, genetics is a scientific field that is young, with a difficult short history. Human genetics was born as a means of measuring people, comparatively, such that the differences between them could be formalized as science, and used to justify segregation and subjugation. The birth of genetics is synonymous with the birth of eugenics, though at the time in the late nineteenth century, that word did not carry the same toxic meaning that it has now. There is no more controversial subject in all of science than race—people are different from each other, and the weight of those differences is something that has caused some of the deepest divisions and cruelest, bloodiest acts in history. As we will see, modern genetics has shown how we continue to get the whole concept of race so spectacularly wrong. Humans love telling stories. We're a species that craves narrative, and more specifically, narrative satisfaction—explanation, a way of making sense of things, and the ineffable complexities of being human—beginnings, middles, and ends. When we started to read the genome, what we wanted to find there were narratives that tidied up the mysteries of history and culture and individual identity, that told us exactly who we were, and why.

Our wishes were not satisfied. The human genome turned out to be far more interesting and complicated than anyone anticipated, including all the geneticists who remain ever more gainfully employed a decade on from the so-called completion of the Human Genome Project. The truth of this complexity and our lack of understanding is struggling to filter down into what we talk about when we talk about genetics. We once spoke of blood and bloodlines as a means of tying us to our ancestors and describing our familial selves. It's no longer in the blood, it is in our genes. DNA has become a byword for destiny, or a seam running through us that seals our fates. But it is not. All scientists think that their field is the one that is least well represented in the media, but I'm a scientist and a writer, and I believe that human genetics stands out above all as one destined to be misunderstood, I think because we are culturally programmed to misunderstand it.

Science is apt to reveal that much of the world is not how we perceive it, whether that is the cosmological, the molecular, the atomic, or the subatomic. These fields are distant or abstract compared with how we talk about families, about inheritance, about race, about intelligence, and about history. The baggage we carry, the subjectivity with which we naturally approach these quintessentially human characteristics is without equal. The gap between what science has revealed and how we talk about families and race is a chasm, because, as we shall see, things are not how we thought they were.

There's plenty of fabrication and mythmaking born of DNA as well. Genetics can certainly tell us who our closest relatives really are, and can reveal so many mysteries of our deep past. But you have far less in common with your ancestors than you may realize, and there are people in your family from whom you have inherited no genes at all, and who therefore have no meaningful genetic link to you, even though in a genealogical sense you are most definitely descended from them. I will show you that despite what you might have read, genetics won't tell you how smart your kids will be, or what sports they should play, or what gender person they might fancy, or how they will die, or why some people commit acts of heinous violence and murder. Just as important as what genetics can tell us is what it can't.

Our DNA is the very thing that has encoded brains sophisticated enough to be capable of asking questions about our own origins, and providing the tools to figure out how our evolution has proceeded. Changes in this strange molecule have accumulated and been recorded over time, waiting patiently for millennia for us to discover how to read it. And now we can. Each chapter in this book tells a different story about history and about genetics, of battles lost and won, of invaders, marauders, murder, migration, agriculture, disease, kings and queens, plague, and plenty of deviant sex.

Above all, you are holding a history book. Some of the stories here are the history of genetics—with all its own convoluted twists and dark past—included to understand how we know what we now are discovering. Many of the stories are tales of nations, populations, a few known through celebrity or inheritance of power, but most are of the anonymous multitudes. We can pick through the bones of individual men, women, and children who through sheer chance died in uncommon circumstances, and turned out to be the people whose lives we would scrutinize forensically because in the preservation of their death they inadvertently gave up their DNA to us.

Biology is the study of what lives and therefore what dies. It's messy—wonderfully, frustratingly so—and imprecise and defies definitions. If you want to start at the beginning, which might seem like a very good place to start, then here is where our troubles begin.

PART ONE

How We Came to Be

1
Horny and mobile

> "There is no beginning, no middle, no end, no suspense, no moral, no causes, no effects. What we love in our books are the depths of many marvelous moments seen all at one time."
>
> Kurt Vonnegut, *Slaughterhouse-Five*

Vonnegut was half right. There is definitely no beginning, and if there is an end, it's not in sight. We are always in the middle, and we are all missing links. Just like there was no absolute point when your life began, there was no moment of creation when our species began, no spark of life, no breath of God into the nostrils of an Adam molded in the red earth, no cracking of a cosmic egg. So it goes. Nothing living is fixed, and all creatures are four dimensional, existing in space, and also through time.

Life is transition: The only things that are truly static are already dead. Your parents had parents, and theirs had parents, and so on, two by two, back through the whole of history, and prehistory. If you keep going back and back, your ancestors will slowly and inevitably become unrecognizable to you, via apes and monkeys, two-legged then quadrupedal, and ratty mammals and brutish beasts on land, and before them in wading sea creatures and fishy swimmers, and worms and weedy sea plants, and around two billion years ago, you don't even need two parents, but just the binary fission of a single cell, one becomes two. Eventually, at the beginning of life on Earth around 4 billion years ago, you're locked in a rock at the bottom of the oceans, inside the hot bubbling tumult of a hydrothermal vent. This geologically slow, incremental change is like a color chart, where pixel-by-pixel white becomes black, whether it's the gap from reptile to mammal, or from four-legged to upright. On occasion there will be a splash of color thrown into the mix, but for the most part, the pathway to your ancestors creeps rather than jerks,* and all of it gray in its depths.

Life on Earth has been continuous in that time, and we are a dot on that gray continuum. Conjure up that image of a hairy monkey-like ape on all fours, to the left of a crouching ape, to the left of a hunched stooping ape, to the left of an upright, modern bearded man-ape like us wielding a flint-tipped spear with his right leg cocked coyly forward to protect us from seeing his immodest instruments of biological transition. This iconic image implies something that we now know is untrue. We just don't know the pathway of the apes that led to us. We know many of the creatures en route, but the map is full of gaps and smears. The second untruth is that there is a direction to our evolution, to our bipedal gait, and our big beefy brains, and our tools and culture. With that arrow we are to infer progress, from simplicity to an inescapable advance into the erect future, an inevitable cognitive revolution of the mind.

Alas, we are no more or less evolved than any creature. Uniqueness is terribly overrated. We're only as unique as every other species, each uniquely evolved to extract the best possible hope for our genes to be passed on into infinity given the present unique circumstances. With all the bones of evolution, and a modern understanding of evolution and genetics, it's impossible to conceive of a

* Creeps or jerks was a joke formulated by the great biologists Stephen Jay Gould and John Turner, concerning the arguments about whether evolution proceeded continuously or with catastrophic game-changing disruption. These two ideas, more formally known as phyletic gradualism and punctuated equilibrium, competed for years. The answer, as so often is the case in science, is, broadly, a bit of both.

twenty-step progress of apes from the left to the right, let alone those neat discrete jumps in five moves. There is no measure of the progress of evolution, and the language we once used, where species were "higher" or "lower," no longer carries any meaning for science.

Charles Darwin used those words,* as was the style of his time, when he outlined the mechanism for the origin of species in 1859. We had scant evidence for other upright apes then, with or without their spears. He had no mechanism for how that modification was passed from generation to generation. Since the end of the nineteenth century we've known the patterns by which characteristics are passed from parent to child. In the 1940s we discovered that DNA was the molecule that transmitted that information down the generations. Since 1953, we've known that the double helix is how DNA is built, giving it the impressive ability to copy itself and allow those copies to build cells just like the ones they came from. And since the 1960s we've known how DNA encodes proteins, and that all life is built of, or by, proteins. Those titans of science, Gregor Mendel, Francis Crick, James Watson, Rosalind Franklin, and Maurice Wilkins, stood on their predecessors' and colleagues' shoulders, and would in turn be the giants from whose shoulders all biologists would see into the future. The unraveling of these mysteries was the great science story of the twentieth century, and by the beginning of the twenty-first the principles of biology were set in place. In cracking the universal genetic code, and unwinding the double helix, we had unveiled a set of simple rules of life. Yet they turned out to be profoundly complex, as we will soon see.

---

* Though he wrote in the margin of one of his notebooks "never say higher or lower," more as an expression of caution for the idea of evolutionary progress. He noted that some barnacles become simpler in chronological evolution. Darwin did love a barnacle.

But Darwin didn't know any of that. When he published his second great work, *The Descent of Man*, in 1871, his primary concern was the question of

> whether man, like every other species, is descended from some pre-existing form . . .

Then, just a handful of Neanderthal remains were known: a skull from Belgium, another from Gibraltar, and a bag of bones from central Germany. As early as 1837, Darwin had sketched out a visionary version of an evolutionary tree in a notebook, showing how one branch of life became two and more, selected by nature in response to the changing environment. How these ancient apes fitted onto the human tree was entirely unknown.

"I think," he scrawled at the top of the page in that notebook, in his inimitably dreadful handwriting, but never finished that thought. What was set in motion in the nineteenth century was the idea that, alongside all animals, we were part of a continuum—a species begotten not created. Nowadays, only the willfully ignorant dismiss the truth that we evolved from earlier ancestors. The images of gigglemug skulls of our long-dead forebears are commonplace, and they become front-page news when a new species is claimed. Dozens of lines of evidence bellow incontrovertibly that we are an ape, with an ape ancestor common to chimpanzees, bonobos, gorillas, and orangutans.

Sometimes people say, as a way of revealing the paucity of the fossil record, that all the specimens of ancient human evolution could be placed on a large table or in a single coffin. That's not true either. We have literally thousands of ancient, hardened bones, found all around the world; many in the nursery of the human story in eastern Africa, many in Europe, and the more we look, the more we find. For Darwin, though, we were effectively alone at the end of a mysterious branch on our family tree.

But for all the sheer grit of the diggers who devote their lives to sitting in dugout caves or dusty ancient riverbeds armed with toothbrushes and tiny picks, there are not nearly enough physical specimens to reveal anything resembling a complete picture when it comes to human evolution; there are individual fossils arranged into groups according to shared characteristics such as the shapes of their brows, the arch of an instep, the cusps on their molars. These were dated according to where they were found, in which layer in the ground, and what other things are found nearby—tools, the shadows of cooking, or traces of hunting.

Or if they're young enough, by the ratio of radioactive carbon atoms that, instead of being replenished via living metabolism, in death are slowly ticking down at a regular rate. It's all good, robust science, contentious as research often is, and frequently fractious, but the analysis of old bones is precise, complex, and highly sophisticated. In the 200 years since the first other human species was discovered, our understanding of how we came to be has undoubtedly increased immeasurably, but our confidence in that path has changed, and continues to evolve. For decades that image of the progress of monkey-ape to ape-man to man-ape has been on display in museums around the world, and in textbooks, a nice line of clear evolution that says "this is how we got here." In Down House, in the English county of Kent, where Darwin beetled away, meticulously drawing up the best idea anyone ever had, you can still buy coffee mugs with that image on it.

When I was young and falling in love with science in the 1980s, the evolutionary trees looked just like that. My father would collect articles for me from *New Scientist* or *Scientific American* showing neat branching diagrams suggesting that one species morphed into another, or one becoming two, with the other gruff ape-men perishing along the way. The picture seemed clearer, the fewer specimens we had. By the end of the twentieth century, more and more human species and specimens had risen from their graves, different enough to blur those nice clean lines, and the branches got fatter, less distinct and more pollarded.

Maybe it's time for us to retire the long-serving metaphor of the evolutionary tree of life, and certainly the picture of apes to ape-men to man-apes. Today, you'd be hard pushed to call it a bush, shrub, or anything arboreal at all. Instead, it is represented in graphical form as more of a set of upside-down dribbly blobs running upward into the pool that is us, streams, rivers, and rivulets, some running into the ocean, others petering out en route (see page 20). An alternative version is to place the specimens in their species clusters on a chart, oldest at the bottom, us the sole survivors at the top, the width showing the geography of where all the bones were found, and you must accept that the lines between them are dotted, meaning hypothetical. If this was a detective story, we have the bodies, but the clues are scant and disconnected. The case is far from settled.

We're utterly fascinated with ourselves, and with some justification. We are just another animal, but we're the only one evolved to have scrutinized our own existence, to look in the mirror and really squint at it. Many, many books have been written about the origin of our species, but this story deals specifically with just those from which we can reconstruct the past, and past relationships, using the newest tool in the paleoanthropologist's shed, and that is DNA. That molecule has revolutionized much of our understanding of human history in unprecedented ways, all in the space of the first few years of this century. It's a field changing so rapidly that researchers have told me of their reluctance to publish new findings for fear of them being superseded not within years or months, but within weeks or even days. Keeping track is not an easy task, as the study of human evolution is mutating into an unremitting revolution. The picture of how we humans came to be what we are is more detailed than ever before, and we still have a long way to go. Before we get to all that, here is a brief, scant overview of the story so far. Let us begin not at the beginning, because there was no beginning, but, somewhat arbitrarily, with two feet.

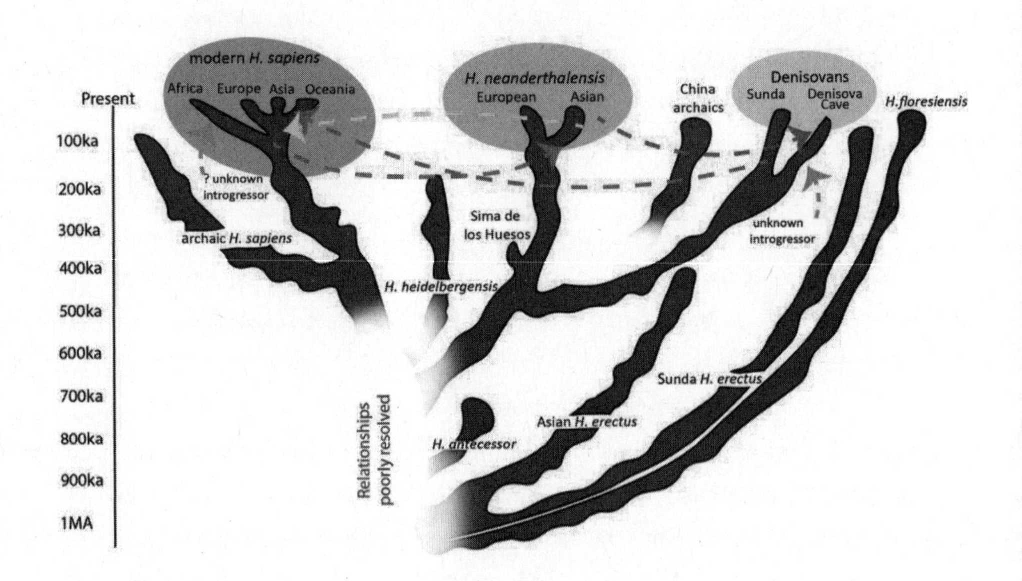

The murky evolutionary shrub of humankind.
Old bones, combined with new analysis of old DNA, has meant that what was once a confident branching tree has been pollarded and pruned and replanted as an unrooted bush. The broad blobs represent individual human species, and the dotted lines are the flow of genes via sex between them. The more we learn, the messier the picture becomes.

Bipedal apes walked the earth at least 4 million years ago. In fact, all apes are capable of two-footed movement, but what concerns us is habitual bipedalism—walking as a primary means of travel. Standing upright was a quintessential step in our own evolution, as it prompted and coincided with a number of anatomical changes, such as the position and shape of the spine, how it connects to the skull, and so on. Why this happened is not agreed upon, and there are plenty of theories: Some focus on an increased efficiency of movement in being upright; others on its being an adaptation to suit a life on the savannah rather than swinging in the trees, or the changing climate of the Rift Valley. The most famous of these early walkers is Lucy, born around 3.2 million years ago. Forty percent of her fossilized skeleton (which is a lot to be preserved for remains that old) was discovered in 1974 by Donald Johanson, and named after the Beatles song "Lucy in the Sky with Diamonds" that was playing back at the researcher's base camp in the Awash Valley in Ethiopia that heady night. Lucy was one of the first members of the species *Australopithecus afarensis* discovered. We cannot say whether her species was a direct ancestor of us. What we can say is that there were many other primates living at this time, and she looks more closely related to us than any other.

The classification of animals is also a frequently unsatisfactory business, but to tell the story of our species we need to dive in and hope for the best. The system that we exclusively use was invented in the eighteenth century by the Swedish naturalist Carl Linnaeus, and gives creatures two Latin names: a genus, and a species.* An

* The ranks genus and species are the last, and thus most specific of the hierarchies of classification, or taxonomy. The standard ranking is Kingdom, Phylum, Class, Order, Family, Genus, Species. I was taught this with the memorable if not kind mnemonic, "kings play cards on fat girls' stomachs," though there are plenty of other versions. "King Philip came over for group sex" may suit your fancies better.

English oak tree is a *Quercus robur.* There's a wasp called *Lalapa lusa*, and a Fijian snail called *Ba humbugi. Enema pan* is the rhinoceros beetle. The common toad is *Bufo bufo*, which might seem a bit lazy as it's Latin for "toad toad," but lots of common animals have this form, including the mollusc *Extra extra*, and our fellow great apes, *Gorilla gorilla.** You may well have a *Felis catus* at home, which at least comprises two different words for "cat."

Lucy's type, *Australopithecus afarensis* roughly translates as "southern apelike thing from Afar." And there are other species of southern apelike things—*sediba*, *anamensis*, and *africanus*. Earlier apes are slotted into genus categories with names like *Sivapithecus* (Shiva's ape, having been found in India), and *Ardipithecus* (ground ape), and *Gigantopithecus* (really big ape).

We are genus: *Homo*; species: *sapiens—Homo sapiens*: the wise man. That's the short version. There's an equivalent in biology to when children write their addresses from street to town to country to continent, hemisphere, solar system, and galaxy. Several classification ranks sit above genus and species to place us precisely in the living universe:

Domain: Eukaryota (complex life)
Kingdom: Animalia (animals)

---

* There is officially a category below species—subspecies—in which different types within a species can be separated. *Gorilla* is the genus, and there are species within that genus: *Gorilla beringei*, informally named the eastern gorilla, has the dubious honor of being named after Friedrich Robert von Beringe, the first person to shoot one. The western gorilla, *Gorilla gorilla* is the most abundant species. But there are also two subspecies of these, one of which is the western lowland gorilla; *Gorilla gorilla gorilla* is its scientific name. It is OK to find this amusing, though taxonomists are deadly serious people.

Most scientists refer to us as "anatomically modern humans." Earlier versions of *Homo sapiens* have been found and documented, the oldest of which, as of June 2017, are the Jebel Irhoud specimens from Morocco, dated to around 300,000 years ago. As an attempt to distinguish between archaic *Homo sapiens* and us, as well as other proposed incarnations, some scientists use the subspecies classification *Homo sapiens sapiens* to denote modern humans.

Phylum: Chordata (animals with a central column, akin to a backbone)
Class: Mammalia (milk producing)
Order: Primates (monkeys, apes, tarsiers, and a few more)
Suborder: Haplorhini (dry-nosed apes)
Family: Hominidae (great apes)

Everything in this book from here on in will be *Homo*. Neanderthals are classified as *Homo neanderthalensis*—the humans from the Neander valley, in Germany; *Homo habilis*—the handy man.

You belong to a strangely exclusive club. Membership of a genus doesn't necessarily show relatedness between the members, but instead shows that members are more similar to each other than they are to organisms not in that genus. This is the best system we have. Species is a definition also riven with problems, but the most accepted form is that two species are defined as distinct when they are incapable of producing fertile offspring together. Zebroids, ligers, mules, hinnies, grolar bears* are all relatively rare, relatively healthy hybrids. But none of them produce fertile offspring of their own. Soon, we shall see why this species definition for humans is not at all adequate.

The current convention is to list around seven species that fit into the genus *Homo*, and I will be referring to them as human. This is not uncontroversial, but one of the key problems of taxonomy is that in trying to name things, we are attempting to describe how things are, and this doesn't necessarily acknowledge the essential temporal nature of life, that evolution is universal, and change over time is the norm. Remember that the subject of evolutionary change is DNA, but classification does not depend upon that.

---

* Zebra with any other equine animal; male lion with a lady tiger; jack donkey with a female horse; jenny donkey with a male horse; polar bear with a grizzly—rare but presumably utterly terrifying.

For the time being, though, let us think of species as distinct groups of animals, who are different enough to be incapable of producing fertile offspring together, and in the genus *Homo*, there have been at least seven.* Ones for whom we have fossil remains from the period starting a million years ago can be called archaic humans, and there are a few. *Homos ergaster*, *heidelbergensis*, *antecessor*, and a few more are all present in different places, and with subtly different anatomical details during this period, and they are all thought to have evolved from earlier *Homo erectus*, the upright human. They did an excellent job of populating the world, but we don't believe so far that they left any DNA for us to recover, and here we are tracking the past with DNA. Most of the others have also not yielded any DNA samples (yet), probably because they are too old, or died in places too hot for it to endure, and so our understanding of our relationship to them is limited to fossils and paleoarchaeology.

The ground of human evolution trembled with the discovery of a tiny woman on the Indonesian island of Flores in 2003. The skeletal remains of a meter-high female, and parts of at least eight other people were unearthed in a cave called Liang Bua. Immediately these miniature humans, classified as *Homo floresiensis*, were referred to as Hobbits, and though their feet are large, there was no evidence of them being hairy. They appear to have lived in this humid cave as recently as 13,000 years ago, which is only a few centuries before the beginning of farming. These tiny people cooked with fire, and probably butchered meat of giant rats and stegodons (a species of tiny elephant) that lived alongside them.

Who were these small humans? The initial reports in the journal *Nature* were clear that their bodies were similar enough to be

---

* There have been dozens more proposed, some more plausible than others, from scientists, sociologists, novelists, and fabulists.

included in the genus *Homo*, but different enough to warrant being a separate species from any other known *Homo*. A vocal minority of scientists criticized this position, and asserted that they were actually like us, but diseased and shrunken through some speculative pathology. Down syndrome, microcephaly, Laron syndrome, and endemic cretinism have all been suggested, but the evidence for these is slight and dicey. Populations on islands often evolve to be very small or big, as the forces of selection might be limited and specific to insular isolation, and indeed the Hobbits shared their island with rodents of unusual size, tiny hippos, and dwarf elephants. All are now extinct, but it seems most likely that *Homo floresiensis* was a separate species of human, probably sharing a common ancestor with us at some point within the last 2 million years, but had shrunk in size due to the pressures of a tropical island life.

But we couldn't get DNA out of the remains of these midgets. The bones were not fossilized, and were soft, like wet cardboard. An attempt to extract DNA was made in 2009, using a tooth, which is hard on the outside, and so offers some protection from the ages. It failed, and their DNA is lost in time, like tears in rain. Maybe the heat and humidity of the tropics over a few millennia were enough to annihilate all of the DNA that might have been skulking in those teeth and decaying bones. It is a shame that this is the case, as there have been heated arguments about the provenance of these people, and DNA would've solved them in a heartbeat. Their island status, the position and limitations of their range, and their physical characteristics suggest that the Hobbits of Flores were not ancestors of ours but distant cousins. Nevertheless, the number of human species who lived into the past 50,000 years had suddenly gone from two to three, and deservedly the Hobbit and its freakish giant and dwarf island cohabitants were famous. Overnight, our planet started to look a bit more like Middle Earth.*, †

Learning to read

And it was going to get more so as DNA reading technology matured. For more than a hundred years, the study of human evolution had been dominated by bones, and a few tools—anatomy

* A brief word about cryptids—proposed, mythological, or vaguely reported species that remain elusive, theoretical, or alien to science. For centuries there have been myths and legends about extant humanlike contemporaries of ours, ape men, yetis in the Himalayas, Sasquatch in Canada, Barmanou in Southeast Asia, Manananggal of the Philippines, the Australian yowie, and Big Foot and many, many others in the rural United States. None of these has ever come close to verification in a way that would slake a scientist's thirst, including such things as evidence of bodies, fossils, hunting or feeding behavior, or even a decent photo, which you might find surprising in a world with literally billions of cameras and now camera phones, and an industry dedicated to finding these elusive monkey-men. When DNA began to become a useful tool in the identification and classification of animals in the 1990s, some studies emerged, and continue to do so to this day, that claim to prove the existence of one of these cryptid ape-men. So far, all of them have been rejected outright by scientists, with a mixture of derision and, in recent high-profile media-courting claims in the UK, weariness. There is the legend of an ape-man on the island of Flores called Ebu Gogo, short and hairy, the females with long pendulous breasts. It is tempting to link the very real remains of the long-extinct *Homo floresiensis* with these legends. But it is far more likely that Ebu Gogo is simply a local version of the myth of orang pendek, a short-statured cryptid story that is common around Sumatra and other islands. Cryptozoology lumbers onward, on the fringes of science and value.

† In the final stages of writing this book, two new studies yet again upturned the Flores story. In April 2016, Thomas Sutikna's paper significantly revised the dates of *Homo floresiensis*. Cave systems are often dynamic geological locations with uneven stratigraphy. Liang Bua is no different, and the most recent date for the Hobbits' occupation of the cave system was amended to be more like 50,000 years ago. In June 2016, a fragment of jaw and some teeth from a family of different, even smaller, Hobbits were found 30 miles east, this time 700,000 years old. They have not yet been named, but there's no doubt much more is to come.

and culture. Reading DNA is really a form of anatomy on a molecular scale; it contains clues to how bones are shaped, and how evolution has shaped them. The invention of the technology to read DNA was primarily born of a desire to understand disease, but it was clear that decoding genomes would illuminate human history too.

Here is a short interlude about how we learned to read DNA. In 1997, in the world of living human genetics, the largest scientific project in history was going full steam ahead. Hundreds of scientists—some former competitors—had effectively teamed up with a common goal: to provide the world with a complete readout of every single letter of DNA in a human being, all 3 billion of them. The tale of the Human Genome Project is told in Chapter 6, but for this story the most important thing is that this was a technological grand scheme designed to make reading the letters of DNA easy and cheap. In doing so, medicine, evolution, and the mysteries of being human would be revolutionized. The ability to read DNA was pioneered by the unassuming English genius Fred Sanger in the late 1970s, using a process that copied the original sequence millions of times. To do that, your ingredients need to include the alphabet you're writing in; DNA consists of only four letters, more formally called nucleotide bases—A, T, C, and G. You also need an enzyme whose job it is simply to copy and link up the bases of DNA, called a polymerase. Throw all these ingredients into a tube and set the temperature right, and the double helix will separate into single strands, which serve as templates to replace the letters that would form the missing strand. You end up with millions of copies of the original template. Each one of the letters of DNA physically links to the one that precedes it and the one that follows, whereas English periods halt any sentence. The polymerase molecule trundles along adding the next letter one at a time like a typewriter copying a line of text. In DNA sequencing, you add not just the correct letter molecules, but also a few that act as periods.

Because so many copies are made during this process, and because the periods are added randomly, what you end up with is a muddle of DNA molecules that stop at

e
ev
eve
ever
every
every s
every si
every sin
every sing
every singl
every single
every single l
every single le
every single let
every single lett
every single lette
every single letter
every single letter.

Sequencing DNA is reconstruction. You make millions of copies that are fragmented at every letter. You then order them by size. DNA is a molecule that carries a negative electrical charge, and this means that if you stick it in some salty water, and put a voltage across that water, the DNA will head toward the positive electrode. The speed at which it migrates is determined by its mass, which is determined by how long that fragment is—a large piece will move more slowly than a small one. So, if instead of putting it in water, you put the DNA in a jellylike gel to slow it down, and run an electric charge across the gel, then the DNA will separate very precisely according to size, just like sieving dirt.

There's one more trick to this technique. There are only four letters of DNA, unlike the English alphabet's cumbersome twenty-six. So, take your original gene, and separate it into four tubes. In each of the tubes you add all the ingredients, but in the first you also add some A bases that halt the chain, ones that add the period, but only when there is an A on the template. In the second, you include everything as well as chain-terminating C bases, and so on in tubes three and four, Ts and Gs. After the reactions are complete you have one tube that contains every fragment of DNA that ends with an A, in the second that ends with a C, the third with a T, and the fourth with a G. If you stick these four solutions in four columns on a gel and apply the current, they will be drawn out and separated, and every position of every letter revealed:

Your A column will look like this (though the letters would merely be smudges on the gel):

*****AA**A*****A******A*A***A*

And the T column:

T*TT**T*T*T*T*T*T****T*T**

The C column:

C**C****C**C*********C*C**C**C

And G:

*G***********G***G*G**********

If you overlay these four together the asterisks become letters, and you get a complete read:

CGTCTAATCATCTGTATGTGTCACATCTAC

When you see TV scientists holding up X-ray sheets covered in dotted black lines in neat columns, that is what they're looking at. It's a sequence of the letters of DNA that sit inside a cell in your body, unreadable for 4 billion years, but now rendered so commonplace that it can be done in minutes for a few dollars. It's an incredibly clever way of reading DNA, and Fred Sanger quite rightly picked up his second Nobel Prize for Chemistry* for inventing it.

During the 1990s, with the Human Genome Project's aim being to sequence 3 billion letters, Sanger's technique was evolved, improved, and automated. There is an account in Chapter 6 as to why this was still a gargantuan task that took years and billions of dollars to complete. When I was a student in the 1990s, I would send off purified samples of short lengths of DNA to a specialist sequencing department and await the results (not on nicely photogenic X-rays, but in computer files) for a few days. Now, most genetics labs have their own sequencers and they churn out megabytes of data in hours. New techniques have been invented that have not completely replaced Sanger sequencing, but are even quicker and cheaper, and were you to begin a career as a geneticist today you would probably never employ this technique. We even have sequencers smaller than a deck of cards that will plug directly

* The first, in 1958, was for determining the amino acid sequence of the insulin protein. He is the only person to have won the chemistry gong twice, and one of only four to have won two Nobels in any category. Fred Sanger died during the writing of this book, but he will be well remembered in his techniques, and in the Wellcome Trust Sanger Institute outside Cambridge, close to where he lived; it is one of the great centers of genome research in the world.

into your laptop via a USB port, so they can be taken out into the field to sequence the genomes of animals and plants in the wild. All these technologies are fueling the revolution in genetics for everyone alive. Since the turn of the century, we have been able to do the same for people who have been deceased for quite some time.

And death shall have no dominion

In a hole in the ground, a man lay extremely dead. He was either left in this tomb by his family or perished right there, with no idea that he was one of the more important people in millions of years. Posthumously—very posthumously—this man did two things: The first is that his emergence out of that cave kick-started the study of ancient humans. It had been his home, we presume, in what we now call Germany, around 40,000 years ago. Kleine Feldhofer Grotte is no longer there; it was discovered but destroyed by quarry miners in the nineteenth century. The entrance stood a few meters above the valley floor, a man-sized squeeze into a rocky room around three by five meters, with a high ceiling. Of the things pulled out by amateur sleuths in the 1850s, and subsequently in excavations of the buried site this century, thousands of artifacts have been found, and the remains of at least three people. Fossilized bones were found in 1856 by the quarry miners—a drink-coaster sized chunk of skull, two femurs, more arm bones than one person requires, and fragments of a shoulder blade and ribs—who handed them over to a local anthropologist.

The remains of this man were not the first (he was probably the third non–*Homo sapiens* human skeleton to have been discovered), but he became what is known as the "type specimen"—the one that defines the species, against which all are subsequently compared. The name of the species is formally attached to the type specimen, so whatever his name was in life, as far as we are concerned, he became known as "Neanderthal 1." With the formal identification of this man, the field of paleoanthropology—the study of ancient humans—truly began.

But they wouldn't let him lie. The second revolutionary thing Neanderthal 1 did occurred another 150 years later. He volunteered his DNA. Up in that cold cave his remains were protected, to a degree, from the elements, from hungry animals, and most significantly from voracious bacteria, all of which would happily eventually destroy all the evidence that he once had lived. Instead, due to his unusual domicile, his bones were left as untouched as someone lying dead for 40,000 years can be, and that meant that he became the first non–*Homo sapiens* human to enter what was, in 1997, a very exclusive club. Tucked inside the slowly decaying cells of what was probably his throwing arm, were the molecules that faithfully carry ancestry from the past into the future.

We modern humans weren't the only ones in the Human Genome Project. Somewhat counterintuitively, there were six species included in the project's primary aims. A genome is much more useful if it can be compared to another, and that includes genomes from other species. So, the original mission of the first creatures to join the genome club aside from us included the most commonly used model organisms—the fly *Drosophila melanogaster*; the rat and the mouse; our closest ape relative, the chimp; and an oddity, the honey bee, for it is a social beast, and almost all members don't get to reproduce at all, but serve their queen with whom they share exactly half their DNA. All of these were due to have their entire genomes read, deciphered, and interrogated over the final years of the twentieth century.

In 1997, using precisely the same techniques being developed for living humans, a Swedish researcher working in Leipzig quietly laid the foundation stones for a new, utterly revolutionary field—paleogenetics. Svante Pääbo had borrowed the right humerus, the bone between the elbow and the shoulder, of Neanderthal 1 from Rheinisches Landesmuseum Bonn. With a precision saw, he sliced an inch-long segment out of the middle, exposing what was once soft vibrant marrow where blood and immune cells would have once been thriving. Bone marrow is the site of a panoply of new cell generation, and that means they are dividing energetically, which means replicating their genetic material boisterously. There lay the first treasure trove of Neanderthal DNA.

DNA is universal in all living species. It's packaged up in different ways, a language organized into books, chapters, origami, and pamphlets. And passed around the generations in different ways too. In animals, DNA is bundled up into chromosomes, huge chunks of double helix, wrapped around itself, and wrapped around little lumpy proteins, which again spiral tight again until they look like those iconic X shapes that we see in textbooks. In most cells, you carry two complete sets of chromosomes, one set inherited from your mother, one from your father, twenty-three pairs in total all neatly stored in the nucleus, the little nut in the center of cells.

Biology is a science of exceptions and endless qualifications, and twenty-two of those pairs are the same as each other (called autosomes), and one of those pairs is not a pair. The pair that is not a pair are the sex chromosomes. I have a Y and an X, whereas women have two Xs. Women get one X from each parent, but men only get their Y chromosomes from their fathers. But, though the Y is important for determining maleness, it's a weedy piece of DNA compared to the others, and makes up very little of the total amount of DNA. The X is the second biggest of all the human chromosomes.

There's another exception to the way DNA is passed from parent to child. All the autosomes and the sex chromosomes never leave the nucleus, a bound space in the center of most cells. But there's a minuscule but terribly important bit of DNA that is not in the nucleus, but instead sits inside the mitochondria—the tiny but powerful energy generation units that all complex life relies on. They were almost certainly acquired around 2 billion years ago when two single-celled organisms fused for mutual benefit. What that meant was that these new cells formed a new branch of life—the eukaryotes—different from everything that had come before, which were small single-celled beings, either bacteria or archaea. These three groups are called domains, and are at the very top of the hierarchy of living things, above the five kingdoms. The three domains are Bacteria, Archaea, and Eukaryota, which is basically everything that is not in the first two categories. The eukaryotes carry inside them a tiny amount of very important DNA that is not stored in the nucleus but bound inside these subcellular power stations. In contrast to the scraggy Y chromosome, mitochondrial DNA (mtDNA) is only passed from mother to child. The sperm swims along with only half the genetic information to make a new person—twenty-two chromosomes and an X (if that child is to be a woman) or a Y—and wheedles its way into the egg, which also carries twenty-two chromosomes and an X, and also the mtDNA of the mother.

Almost all—more than 97 percent—of your DNA is carried in the twenty-two pairs of autosomes and the X, and all this genetic information is inherited from both parents in a roughly equal manner. Each of the autosomes is a unique combination of the pair of chromosomes that your mother or father inherited from their mother and father. When a sperm or egg is being forged in your father's testes or in the ovaries of your mother* the two matching chromosomes line up and shuffle. Imagine lining up two suits of cards, hearts and clubs, in a row, and then swapping some of the

---

* This process occurs in utero. The egg that made you was made inside your mother's ovaries while she was inside her mother. Your DNA was forged inside your grandmother.

same numbers with each other. The result would be two complete suits, all the right numbers in the right order, but a mix of hearts and clubs. This is what the chromosomes do in making sex cells. But instead of ace to king each chromosome has millions of possible swaps. So the result is, for each of the twenty-two autosomes, a new combination. It is this process, called "recombination," that guarantees that your precise genetic makeup is unique to you, for all time.

Mitochondria and the Y don't do this though. One comes from your mother, which came from her mother, and from her mother, and all the way back only on your maternal side, and the Y exactly the same on your paternal side. For those in pursuit of ancestry, these make for interesting tools, and ones that have been the focus of many studies, not least because they have been historically the easiest, smallest, and therefore the first chunks of DNA available for ancestral scrutiny. Mitochondria exist in their millions in the busy milieu inside cells, and so the chances of their surviving the onslaught of time is higher. The autosomes and sex chromosomes exist as one complete set kept exclusively inside the nucleus—the cell's central office. So compared to the nuclear DNA, there's stacks of it, millions of identical copies all much more readily available for analysis. Both mtDNA and Y will make frequent appearances in these pages; not just because they are informative, but also because their value is sometimes overstated in the hunt for ancestry.

The Neanderthals were a people who lived all over western Europe from the easternmost tip of Spain, to the caves of north Wales, into the mountains of central Asia, and as far south as Israel. The oldest true Neanderthal bones we've found are 300,000 years old, and we haven't discovered any younger than 30,000. That is a reasonable longevity for a human species. *Homo erectus*, an earlier upright ape, spread all over the world from an exodus out of Africa that began 1.9

million years ago. But the Neanderthals still clocked up a longer run than we have so far. We anatomically modern humans are generally thought to have evolved primarily in eastern Africa around 200,000 years ago, and emerged out of Africa in our own exodus sometime in the last 100,000 years. This number inches up every few years, as more specimens are found. A discovery in October 2015, from the Fuyan cave in the Daoxian region in southern China, dug up forty-seven modern teeth at least 80,000 years old, and it's not unreasonable to presume that the owners of those teeth took some tens of thousands of years to get that far east from the motherland.

According to traditional paleoanthropology based on bones, by the time *Homo sapiens* reached Europe, probably around 60,000 years ago, the Neanderthals were already there and well established, albeit in small communities. But with DNA as evidence, these dates are due for serious revision, as we shall see later in this chapter.

Nevertheless, Neanderthal anatomy clearly shows that they were visibly different from the interlopers. Brain capacity is one of the key measures in paleoanthropology, and the Neanderthals had bigger ones than us; a modern man's averages about 1.4 liters by volume, women's being a little smaller. The Neanderthals range between 1.2 and 1.7 liters. That cranial capacity is not something we can specifically correlate with intellectual capabilities, though in general, in apes, bigger brains mean more sophistication.

They were shorter and stockier than us, thicker set, barrel chested, with broader noses and clunkier brows. It's surely for these simple physical reasons that they have a pretty bad rep. In common parlance they are synonymous with and stigmatized as brutish cavemen, grunting oafs, and *neanderthal* acts as a byword for low-brow dimwit thuggery. When the classification of the early specimens was being wrangled in the nineteenth century, the great German biologist Ernst Haeckel suggested one specimen be known as *Homo stupidus*.

Nothing indicates that the Neanderthals were anything of the sort, nor significantly different from their *Homo sapiens* contemporaries. They hunted, butchered, and cooked large prey. In the last 100,000 years there is some evidence that they sewed, made clothes and jewelry, and these artifacts predate the arrival of anatomically modern humans, meaning that the Neanderthals developed these skills themselves, rather than learning them from the new kids on the block. Recently researchers have claimed that hand paintings in a cave in Nerja on the pebbly Mediterranean coast of Spain were theirs and not ours. Some have argued that the presence of pollen in graves in Shanidar in Iraq and in southern France* were traces from flowers left there as part of ritual burials, though this remains controversial.

Because of the paucity of remains, the evolutionary relationship between *Homo neanderthalensis* and *Homo sapiens* has long been disputed. The full gamut of suggestions has been made over the years, from their being the direct ancestors of modern Europeans, to their existence on a completely different bough of the evolutionary tree, who left no extant descendants. The last common ancestor of us and them is thought to have existed around 600,000 years before today.

Svante Pääbo's digging within Neanderthal 1's arm bone was the first step in answering this. They extracted 0.4g of matter—the weight of a decent pinch of salt—from the section of precision-butchered bone, and from it pulled fragments of mtDNA. This was, in 1997, the most ancient DNA yet recovered. Much of that first study was devoted to showing that it was possible, and that the DNA extracted was not contamination.

---

* Bones discovered in La Chapelle-aux-Saints in 1908 were the ones that gave rise to the stereotype of a hunched caveman oaf. In the 1980s, much more forensic analysis by Eric Trinkaus showed that this was a forty-year-old man hunched because of osteoarthritis, not because that's how they all stood.

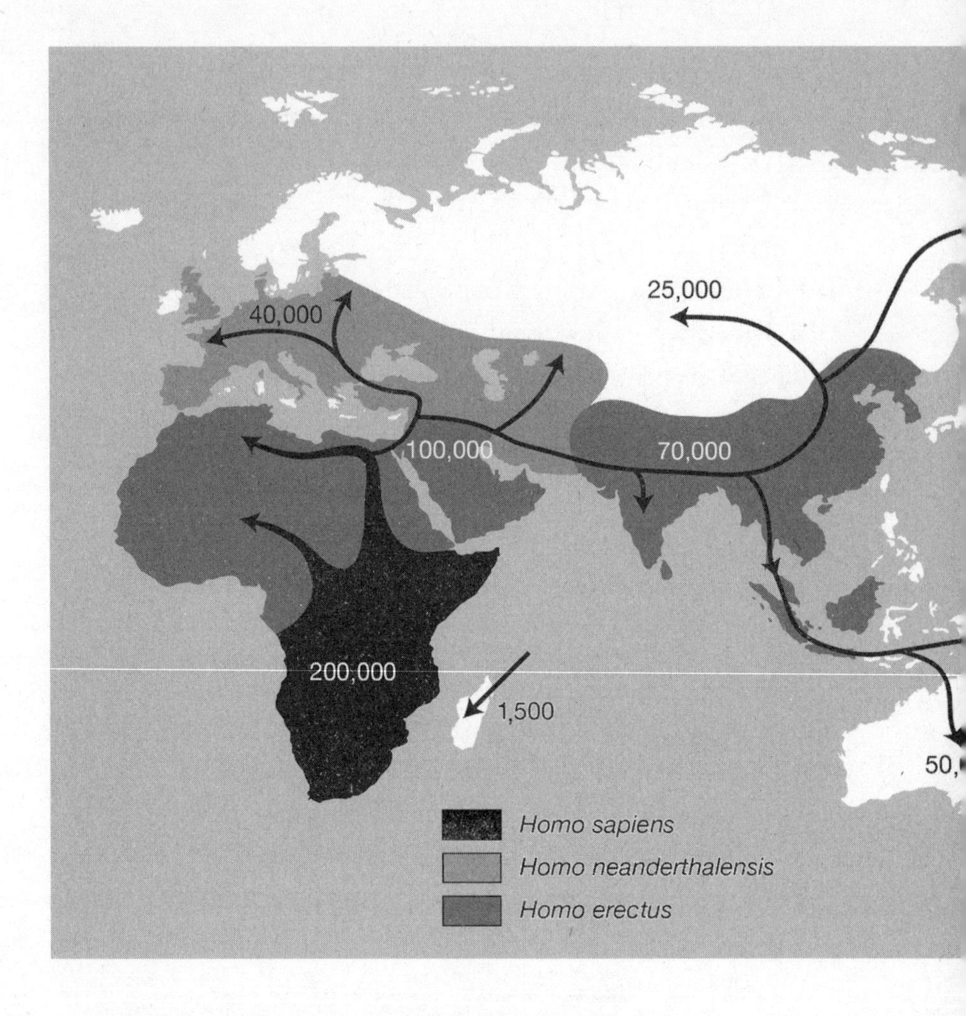

The migration of Homo sapiens out of Africa.
Anatomically modern humans began their tenure on Earth primarily in eastern Africa around 200,000 years ago, though more archaic *Homo sapiens* have been found as far away in time and space as 300,000 years ago in Morocco. Our ancestors had begun to trickle out of Africa at least 100,000 years ago. They met Neanderthals in Europe, and other human species en route, and according to our DNA, bred with many of them.

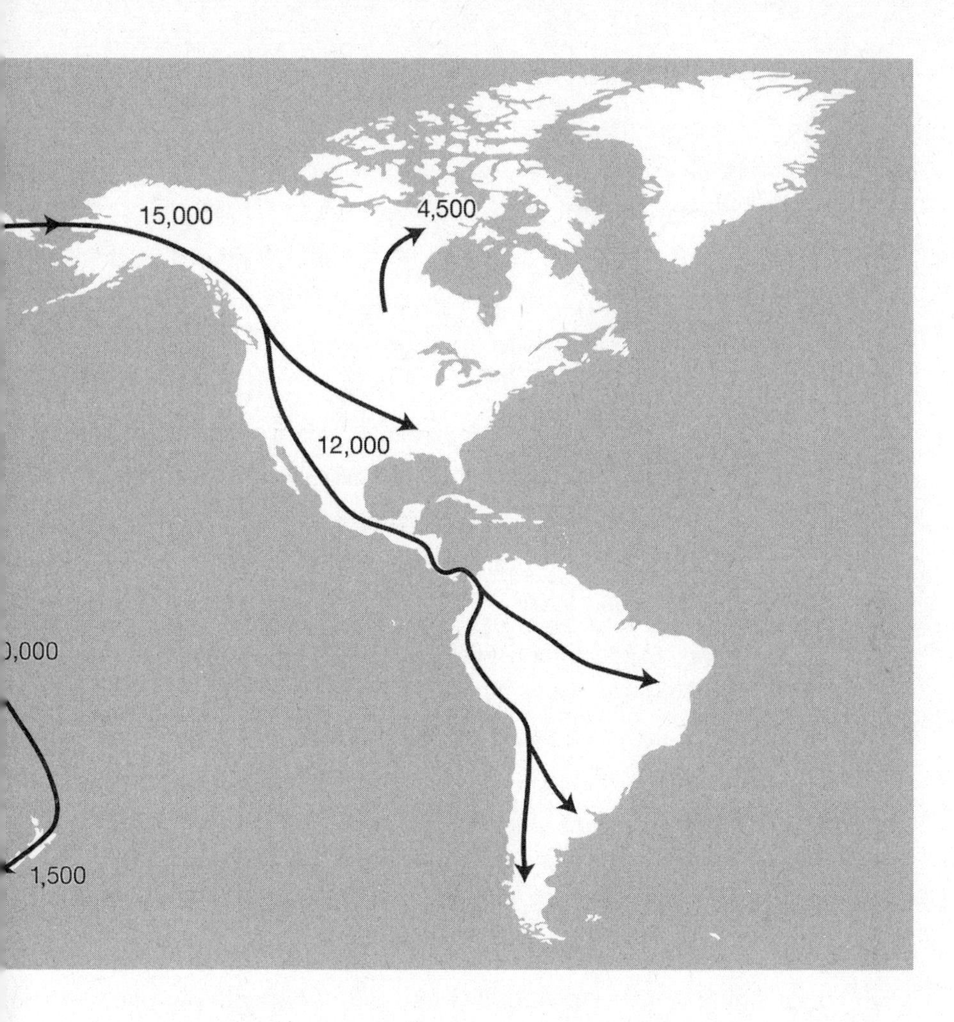

As with all ancient DNA, much of it is not human in origin. Some contamination might be the DNA of organisms that have inveigled their way into the rotting corpse looking for a feed. Some might have been the organisms—mostly harmless or even beneficial bacteria—that live in and on us and outnumber our own human cells. And some might be malicious pathogens, maybe even the thing that killed them.

Jurassic Park and its lesser sequels had become record-breaking hits in the 1990s, and the idea of DNA recovered from long-dead species was very clearly in our cultural consciousness. The reality, as ever, was rather short of the stories in the movies. The DNA was chunked up into short fragments, and all profoundly damaged, as if recovering frayed tatters from a decrepit book. These bones were only 40,000 years old, in contrast to the more than 65-million-year-old resurrected dinosaurs* of *Jurassic Park*, and already not in a terribly good state. The fact that Pääbo and his team got anything out is testament to the newfound skills of geneticists that were emerging in the shadow of the Human Genome Project. This was a baby step into deciphering and reconstructing the past in an entirely unprecedented way.

The first thing it said was that the DNA plucked from the Neanderthal man was different from all modern human mtDNA. The sequence of the fragments of DNA analyzed is different enough to say with some certainty that this part of their genome had separated from the lineages that led to all modern humans well before a common ancestor of all modern humans. DNA changes over time in a relatively predictable manner, like a slowly ticking clock, and so by taking two sequences that are similar but different, we can estimate how long ago they diverged. This technique is not perfect, but it has value in broad terms. In the case of this first study of Neanderthal DNA, the age of the divergence between us and them was put at between 550,000 and 690,000 years ago. These were both reassuring for the traditional forms of human evolution: Neanderthals were not us, and have not been us for an age pretty much in alignment with what the paleontology and archaeology said. The status quo had not been upended by this technological

* Actually they were almost all Cretaceous animals in that film. But *Cretaceous Park* is probably trickier to pronounce.

feat. In unlocking this door to the past though, over the next decade, everything would change.

The revolution accelerated alarmingly, but always the brake was the profound skill required in the process. Extracting ancient DNA is not an easy thing to do, and the volume of ancient DNA studies is testament to truly expert skill. Gene sequencing from living cells these days is easy as pie, and anyone can do it with a couple of day's training and the right equipment: It's the analysis, the number crunching that takes real expertise. Compared to its living equivalent, ancient DNA is a fragile wisp, and because of the delicacies involved, reading the genes of the long dead will never be normalized in such a way that anyone can do it.

But, as with the Human Genome Project, part of the deal is that when you sequence these chunks of DNA, they become public. These ancient genomes are published as databases, free for all to plunder. Geneticists don't have to go near a fossilized bone or a dank cave nowadays to quiz the genetics of our millennia-dead ancestors. You just need the Internet. The first few extractions did require trailblazing development of new techniques in preserving and analyzing ancient DNA, because they were limited in what they could actually extract. In 2006, another team successfully pulled out DNA from a 38,000-year-old Croatian Neanderthal, and used it to answer some old questions. Two papers, with some shared authors, were published in the two top research journals, *Science* and *Nature*, in the same week, and the results were robustly similar, and subtly different. The key finding was that the sequences of DNA generated implied that the human that led to us diverged from those who led to the Neanderthals around 500,000 years ago. One hinted at what was to come: that there might have been a touch of interbreeding at a later stage, a tease of surprising sexual dalliances. The other said there was none.

❧

Then in 2010 came the complete Neanderthal genome. Svante Pääbo's team had radically improved their techniques for extracting DNA from ancient bones. Using dust, chips, and flakes drilled out of fossils, they assembled a full draft—rough though it was—of the complete DNA from a Neanderthal.

Let's take a moment to revel in this. The speed of progress has been truly breathtaking. Our contemporary genomes had only been deciphered in any near-complete form in 2001 (and, as described in Chapter 6, really properly in 2003), and yet within a few years we had completed a draft genome of an extinct human species, from bones untouched for tens of thousands of years. Pääbo and his colleagues had invented a time machine.

What were they like? What genes are and what they tell us about people are very closely related, but not, in almost all cases, definitive. This is a seam that will run throughout this book, confronting and dispelling the culturally ubiquitous idea that genes are fate, and a certain type of any one gene will determine exactly what an individual is like. That this is a fallacy is universally known among geneticists, yet it is still an idea that carries a lot of cultural significance, fueled frequently by the media and an ultra-simplistic understanding of the absurd complexities of human biology. Knowing the gene sequence in an individual provides some limited information, unless they have one of a relatively small set of genes that have a very significant effect. This is discussed in much greater detail later on, but for here it is, for better or worse, the main way we can analyze the genes of the long dead.

A popular exam question for undergraduates studying paleoanthropology is "Did Neanderthals speak?" The correct answer, to be spread over 3,000 words of supporting anatomical evidence please, is that Neanderthals were capable of speech, very probably. The structure of their throats is not dissimilar to our own and, in particular, the discovery of a hyoid bone in the Kebara Cave in Israel in 1989 indicated that their capability of speech must have been similar to our own. The hyoid is a horseshoe-shaped bone where your neck meets the underside of your chin; feel it with your thumb and forefinger (where you might throttle someone) and then swallow. Its structure in humans is uniquely ridged and supported by delicately tuned muscles that connect up in all directions, to the tongue, to the mouth floor under the tongue, to the larynx, the pharynx, and the epiglottis, twelve in total. That's a lot of very fine muscle for a bone so small, and indicates that it's quite specially adapted for something that we do uniquely. The Kebara hyoid has been the subject of many detailed microanatomical studies because of its potential role in speech. And the answer is still the same: Neanderthals probably had the capacity for speech, like us.

From a neurological point of view, we can also make not much progress. The capacity for speech occupies a large chunk of our brains, as you would expect. Although something as complicated as speech engages many parts of the brain, the key zone is known as Broca's area, named after the nineteenth-century French neuroanatomist who treated two patients who had lost the ability to speak after injury to this region. This scale of neuroanatomy is not that useful to us in answering the question of Neanderthal communications, as it's a largish piece of brain, a hunk of meat that you might find in a stew. It's there in the other great apes too, though they cannot speak. So, given that Neanderthals had larger brains on average than ourselves, we can reasonably assume that their Broca's area was present and correct as with all great apes.

You might think the answer would spring up out of the past by using the new genetics. There is a gene, much studied and much lauded, that is inextricably bound to speech. It's called FOXP2, and although we don't really have a great understanding of what it does in the body, it is clear that it is essential for the type of verbal communication and dexterity that we find so trivially easy, compared to our nearest ape relatives, and indeed all other life on Earth. One of the key ways we know how genes work is to look at what happens when they go wrong. We do this deliberately in experimental animals, precisely or randomly disrupting genes to see what happens. For obvious reasons, we don't do that in people, but the equivalent is to study the genetics of disease and disorders. Several mutations in FOXP2 have been identified since the gene was formally described in 2001 by Simon Fisher and his team at Oxford University. The identification relied on a single family (known only by the initials *KE*) that had been referred to Great Ormond Street Hospital for Children and its research wing (and my alma mater), the Institute of Child Health at University College London. The KE family had a heritable form of developmental verbal dyspraxia, where children present a phalanx of vowel and consonant problems. The passage of inheritance in this family, sixteen members profoundly affected, led to the identification of a region on the seventh chromosome that was different from normal, and subsequent studies showed that they had a spelling mistake in this particular gene.* Since then much has been made of FOXP2, including popular press fanfares about its being "the grammar gene" or "the language gene." It is neither, because language and speech are complex behaviors controlled by many genetic factors, and not merely a single gene. But clearly it has a major role in how we communicate.† Furthermore, when we look at the equivalent versions of FOXP2 in other animals, we see that it plays a big part in vocalizations across

---

* There are many types of mutation in DNA that cause problems or evolution. All alter the production of the protein that is encoded by that stretch of DNA. Some delete single letters: Genome becomes Gnome. Some add a period before the sentence is finishe. A missense mutation, as seen in the KE family, introduces ectopic punctuation, so the protein becomes mangled: "too late" becomes "tool ate." If these mutations do not cause serious problems or death, then they can be the source of variation, which is the fuel for evolution.

† This phenomenon, where a single gene becomes known as the single cause of any characteristic, is interrogated in Chapter 5.

the animal kingdom. Male zebra finches cannot direct their songs to females with a suppressed FoxP2, and mice pups that have had their Foxp2* deleted cannot make the ultrasonic peeps that are an essential part of communicating with their mothers.

FOXP2 is certainly essential for speech. The 2006 analysis of the Neanderthal genome revealed that they had exactly the same type of FOXP2 gene as us, and different from chimpanzees. The differences are subtle but clearly important. There's only two changes between the protein sequence of FOXP2 in chimps and us, and we can talk and they cannot.

Can we say the same about Neanderthals? The basic answer to that undergraduate exam question "Did Neanderthals speak?" remains the same as it did when I answered it in 1995, though, if you are an undergraduate looking for an easy pass, the content of your essay today should be entirely different. *Plus ça change, plus c'est la même chose.* Neanderthals very probably had the capacity for speech. Until we genuinely do invent time travel, it is going to be impossible to prove.

If that story doesn't offer a satisfying conclusion (and the study of old bones frequently does not), here's another one of the senses for which we can offer a solid, perfect answer: smell. Before we get to a singular sweet conclusion though, here is a customary scientific disclaimer: We don't really understand how smell works. In our noses there are cells that are triggered when particular molecules float onto them. We capture airborne whiffs with proteins called olfactory receptors, which sit astride the membrane of the neurons in our noses that wire directly to the brain. They're very similar to the proteins in our rods and cones in our retina that trigger vision,

* The keen-eyed might have noticed that there are three versions of FOXP2 written on this page. This is because, rather vexingly, geneticists use different versions of capitalization in different species: FOXP2 is human, Foxp2 is mouse, FoxP2 is zebra finch. I have no idea why.

but instead of being stimulated by photons of light, they fire their signal after trapping odorous chemicals, and thus begins the perception of smell. Unlike with vision, there are many types of olfactory receptors and each one seems to capture a range of smelly molecules. To complicate matters even more, each smelly molecule seems to stimulate multiple receptors.

As with all proteins, olfactory receptors are encoded in genes, written in DNA and set in our genomes. We have around 400 olfactory genes, ones that relate to smells, and these combine in myriad ways to create the rich smellscape that we enjoy. How these combine to create complex perception remains something of a mystery. But there are some exceptions, and a neat study in 2015 focused on one of these, presumably so we didn't all get too despondent that human genetics is all a bit inscrutable. OR7D4 is an olfactory gene, but mercifully has a very direct relationship with how we smell, as it only detects one molecule. Different people have different variants of this gene, and unusually, they correlate pretty precisely with what we think of the smell of a chemical called androstenone. We don't really know what this steroid does in or for people, if anything at all, though it is found in sweat. But if you're a pig, then it's the main way you get laid.

Male pigs and boar produce it in their saliva, and it's the principal component in "boar taint," the smell of uncastrated males. When a female gets a sniff, she may well adopt a mating stance, if the mood takes her. This, as far as I am aware, does not happen in humans, but some of us can smell androstenone and others can't. To many, it honks like stale urine;* to some it is sweet; and to me, well, I can't smell it at all. These perceptions are determined by what allele of the OR7D4 gene you carry on your chromosome 19.

---

* To the Italian tourist who vomited after I asked her to smell a tube of androstenone for a BBC television series in 2010 in Covent Garden Flower Market in London, I apologize.

OR7D4 is a perfectly ordinary gene. It's one of roughly 20,000 that we carry around in our cells, just under a thousand letters—nucleotides—long, which is a very typical length for many genes. Variations in a few of these letters determine what you think when some boar taint wafts up your nose, and these have a very direct correlation with a very real and visceral sense, whereas most other smell receptor genes are more complex and nuanced in how the sequence in the gene relates to what we actually perceive.

I asked Matthew Cobb, one of the scientists behind this ancient whiff, why they had asked the Neanderthals this question, and the answer was simple: "Because we can." I like that answer. The real purpose of the paper was not an insight into the smellscape of the long dead, but to see how the various versions of the gene are distributed across the world today. In doing so, Cobb and his colleagues speculate that the variations might share a distribution pattern with pig farmers over recent human evolution; later we'll see how milk farming had a similar effect in the same way. But this slight sidestep into the deep past typifies how easy contemporary genetics renders quizzing the ancients. This simple, not particularly important gene, says something almost magical. Our desire to reconstruct the past has taken a new turn into the realms of the senses. We now know that if a Neanderthal came across a frisky boar, he or she would be revolted, unlike me, who remains utterly unmoved.

One of the analyses of the Neanderthal genome looked in on a gene called MC1R. This is one that encodes a pigment in skin and hair. There's a rare version of this gene that some people carry, and if they have the same rare version on their sixteenth chromosome inherited from both parents, they will have red hair. There are a few alleles of MC1R that result in being pale-skinned and ginger,*

* More on red hair and MC1R in Chapter 2.

but one of the most common ones is a single letter change, from G to C, three quarters of the way along the gene. Looking at the Neanderthal MC1R reveals some of them had a different mutation, one that we haven't seen in living humans, but which may have meant that some of them had reddish hair and pale skin too. There's no hair left from those people, and pigmentation doesn't survive well over time. So we test it by taking the version of the gene from the Neanderthals and inserting it into a bacteria or other small organism that we can genetically manipulate, and see what it makes. This system doesn't grow a hair, but simply determines the melanin type that will fill the melanosomes, that will color the skin and hair. The results were equivocal. So it goes. Were they ginger? Maybe, but if they were, they weren't ginger like us.

A continuing theme of this book is the limits to what genes can tell us about us. I could list a few more of the genes that have been identified in Neanderthals as being the same as, similar to, or noticeably different from our own, but the relationship between how genes are spelled in our genomes (the genotype) and how they manifest themselves as proteins and ultimately showable characteristics (phenotype) is frequently not very clear. Genes with clunky abbreviations are there in the Neanderthal genome as they are in our own, and we can speculate about precisely what they do. One of the grand endeavors of twenty-first-century genetics has been to test the function of genes by manipulating their equivalents in mice and seeing what happens, or by identifying human diseases in which those genes are mangled. We could speculate about SRGAP2, which has been linked with intelligence in some studies. We (and Neanderthals) have more copies of this gene than chimps, and probably more connections in our neurons as a result. Or we could talk about HACNS1, which is involved in hand development before birth, and is very different from the version that chimps have, whose manual dexterity is not as sophisticated and articulated as our own. But we don't really understand what HACNS1 does during our own development, and therefore its role in the evolution of our handy skills. With all of the genes of Neanderthals potentially known, the truth remains that in most cases, knowing precisely how a gene is spelled is not enough to say precisely what it does.

This is not to undermine a century of genetics. I've picked out a few things about Neanderthals that can be determined by looking at individual genes, which are of interest—mostly trivial interest—hair color and smell. What turns me on is the broader arc of prehistory. Genetics turned into genomics when we began looking at all the DNA in an organism rather than just a handful of genes. Sweeping the whole Neanderthal genome is much more interesting than inspecting individual genes because it tells us about us. The question of our relationship with Neanderthals has been refined with genetics, in terms of our shared ancestors; our lineage moved away from theirs around half a million years ago. But what DNA analysis revealed more categorically than anything else was that we had sex with them, repeatedly, probably as soon as these two peoples met, and every time afterward.

So what happened? Humans are both horny and mobile. The language we use feels deceptive in these terms, at least in the timescales we're referring to. When we say humans migrated out of Africa, as our ancestors surely did, it sounds a little like they packed their bags and headed north to the Promised Land. The whole basis of current thinking about the origin of us is referred to as the Out of Africa hypothesis, defined by our migration away from the first site of anatomically modern humans. The timescales are not really known precisely, other than to say that they were over thousands of years. Our *Homo sapiens* ancestors inched into Europe around 60,000 years ago, and that story is told in Chapter 2. They didn't turn up overnight with suitcases though. The spread of small groups or tribes expanded in all directions, including fundamentally away from where their ancestors were. That's the best we can say. The first studies showed that there were at least five of those euphemistic gene flow events, but again, that doesn't mean that five individuals had sex and produced offspring that lived their lives, and on into the distant future. It means that populations, tribes, interbred and shared their DNA across those populations.

All the Neanderthal DNA sequences are available as online databases, and nowadays the latest sequencing technologies mean that everyone can have their genome scanned (though not fully sequenced) and analyzed for many different things. 23andMe is one such company, and I had my genome parsed with them, the results of which are discussed in more detail later on. But one of the things that emerges out of these personal genomics is what Neanderthal DNA you carry. For me, a solid 2.7 percent of my total DNA is drawn from these people (which rather uninterestingly, according to their data, is exactly average for most Europeans; academic results suggest that this is an overestimate, and the proportion is lower in Europeans). Three billion letters of DNA make up my genome, and based on the 23andMe data, around 81 million of those come from Neanderthals, spread in chunks of varying size across my twenty-three pairs of chromosomes. Six whole human chromosomes have less DNA than that, including the Y, which makes me manly. Admittedly, this contribution is not all in one lump, nor is its influence felt in a singular way. But it is there in me.

For a century the Neanderthal people have worn a stigma of hunched brutish grunting cavemen. Facial reconstructions of their skulls show them to look not exactly like us, and not exactly pretty. But beauty is a very subjective matter, and just because you don't fancy them doesn't mean that your ancestors didn't. They definitely had sex with them.

I carry Neanderthal DNA. Therefore, Neanderthals were my ancestors. If you carry their DNA, then they were your ancestors. If you are broadly of European descent, then it is almost certain that you also carry around Neanderthal DNA. This idea is called introgression—the introduction of DNA from a separate group over repeated familial backcrosses. It is a form of fusing some bits of DNA from distinct populations. Since their discovery, Neanderthals were referred to as cousins, or close relatives. It seems clear to me, using genetics as my ally, that Neanderthals categorically were also our ancestors.* This is one of the key concepts of DNA ancestry, the one that confuses all the nice clean hypothetical branches of a family tree. It's called admixture—the mixing of genes from populations who were previously separated.

We carry their DNA. It's not present in all human populations: Most Africans have very little; some eastern Asians have more than Europeans. In the longer term, "gene flow events" is how admixture occurs. But let's be functionally crude, we're talking about sex. Anatomically modern humans had sex with anatomically Neanderthal humans on many occasions in our history.

We learned from Croatian Neanderthal bones that we had interbred around 60,000 years ago, a time when it was presumed *Homo sapiens* had first reached Eurasia. As soon as we met, we mated. Romanian bones showed that it happened again around 40,000 years ago. It was beginning to look as though whenever our ancestors encountered *Homo neanderthalensis*, they got it on.

A female Neanderthal who died 50,000 years ago in the Altai Mountains in Siberia had joined the genome club in 2014, her toe

---

* I wish to acknowledge Lara Cassidy, a researcher at Trinity College Dublin. I hadn't thought about Neanderthals and the problems of species definitions very hard until she challenged me in a public lecture. She was right and I was wrong, and it made me think, so I thank her for that.

bone being the source of the most detailed DNA yet recovered for her kind of human. But two years later, in February 2016, more meticulous analysis of her genome yet again upended what we had thought was true. It showed that she carried some modern human DNA, and by comparing it to others, we can put a date on when that introgression happened. It had occurred in one of her ancestors around 50,000 years before she was born. We don't know who it came from, but whoever they were, they were like us, and they were a long way in time and space from when and where we thought the African diaspora of *Homo sapiens* had reached. It may be that these people represented the first wave of African emigrants, perhaps a kind of scout party pioneering east, tens of thousands of years before we had thought any of us had left the motherland. The Out of Africa hypothesis remains completely intact in principle, but the dates and the overall flow have profoundly changed with evidence provided by ancient DNA.

Whenever they encountered each other, the genes flowed. When we talk of gene flow events, the word *events* is potentially misleading, just as the word *migration* is. Given the widespread distribution of Neanderthal DNA in populations across Europe today, it is unlikely that this arrived in our gene pool after one of your ancestors mated with another probably shorter, stockier one. These events refer to interbreeding on a population scale, and the question of why this admixture introgression occurred is an interesting one.

When eggs and sperm are made, the gene shuffle of recombination is random, so the genes that an individual will acquire as the result of an interspecies hookup will be just the luck of the draw. As they spread through the population as a result of that individual breeding, natural selection can cast its hand over the usefulness of those genes. In autumn 2015, a sweep of studies analyzed how well evolution received the various genomic acquisitions from Neanderthals. Because DNA tends to be inherited in chunks, we can learn how useful bits of genome are by the size of chunks that are shared. A version of a gene that is definitely useful and therefore likely to be selected by nature over time may carry other bits of flanking DNA with it as it passes through the generations, like a peloton in a bike race being driven forward by the presence of a top rider, bringing all the pack with them. Or the genome may slowly ditch them if they are harmful. By comparing sections of Neanderthal DNA with sections of DNA in modern humans that we think have come from Neanderthals, we can build up a very finely tuned model of the success, from an evolutionary point of view, of these hybridizations. When Graham Coop and his colleagues did this, they found that our genomes are slowly purging themselves of Neanderthal DNA, which suggests that these matings were not to our advantage, but not massively disadvantageous. Our DNA around Neanderthal chunks is undergoing weak negative selection, the peloton slowly decelerating as a result of a weak rider. It might have to do with population size. It's likely that their numbers were always low; mtDNA taken from various Neanderthal bones is all pretty similar, which indicates low genetic diversity, which suggests a small breeding population, maybe just a few thousand. We think that when these meetings were taking place, we were much more populous than them. Once their DNA entered ours, even if it were slightly not to our advantage, then it could be swamped out by a much bigger gene pool. While it was in them, it could be perpetuated as there was no notable better DNA with which to compare. A second paper, by Kelley Harris and Rasmus Nielsen, found the same thing, and also that the selection against introgression was not strong enough to suggest an inherent barrier to reproduction, as you might see in distantly related species.

Another quirky detail that helps us understand these dalliances a tiny bit more emerges from this number crunching. The amount of introgression from Neanderthals is proportionally lower on the modern X than on the rest of the chromosomes. X chromosomes are only passed on by males half of the time because we also have a Y, but all of the time by women, who have two Xs. The observation that there is less Neanderthal DNA on our Xs implies that the first encounters we had with them that resulted in procreation were male Neanderthals with female *Homo sapiens*.

What did the Neanderthals ever do for us? Not that much. We can't tell why we have been slowly rejecting their DNA for thousands of generations. One of the important lessons here is that it demonstrates the speed of evolution, or rather, its breathtakingly slow burn. Any seriously deleterious effects would have been wiped out immediately, and maybe lost to time. However, the fact that their DNA mixed with ours, and is only being selected out of thousands of generations, indicates that some form of hybrid incompatibility was not apparent. These analyses are extreme fine-tuning, only visible through the microscope of statistical inspection of DNA sequences. And they most certainly do not suggest that Neanderthals were a separate species.

We can't say how this interbreeding happened. Was it forced? Or mutually consensual? We don't know. We first met in Siberia 100,000 years ago. We coexisted in the main body of Europe for more than 5,000 years, which is by comparison almost as long as written human history. If you consider our understanding of the last 5,000 years of history, during which time much of it has been explicitly documented, and then consider what has been recorded by proxy in the prehistory of Europe, then you see the scale of the problems of reconstructing the deep past. Our relationship with the Neanderthals has been scrutinized for decades and we know that we lived and bred with them. But some archaeological research suggests we may also have hunted and eaten others. The distribution of the Neanderthal people was widespread, and we probably encountered them all over Eurasia, but it may be that their low numbers resulted in a bottleneck, a lack of genetic diversity that renders a population less healthy overall. It may be that we brought with us diseases that they had not evolved to counter. Our existence ultimately subsumed theirs. The Neanderthals were a proto-species, an embryonic light that flickered in evolutionary time, but was not strong enough to stand across epochs. Whatever the reason for their dwindling from not many to none, we carry their genes, and their immortality will be as enduring as our own.

A tooth and a fingertip

The Altai Mountains loom out of the ground near the Russian borders with China and Mongolia, and they are icy cold. This land is harsh. There's a cave in this hinterland of Siberia, called Denisova, named after an eighteenth-century eremite called Denis who lived there. Due to the brutal weather it's inaccessible for much of the year. Modern human and Neanderthal remains have been recovered from Denis' cave over the forty years it has been explored, as well as dozens of species of animals, from lions and hyenas to woolly rhinos and, as befits the Russian motherland, a lot of bears. Soviet researchers had pulled out over 50,000 artifacts from this cave right up to the Middle Ages, indicating that it had been occupied in some form for more than 230,000 years. Siberia is one of the least populous areas on Earth; today it averages three people per square kilometer. For Denis' cave to be such a regular haunt for hundreds of thousands of years is not usual. You can see why though. Despite the gritty weather, it's a highly desirable residence; a waterfront property overlooking a picturesque river, the rustic estate boasts a wide rectangular south-facing entrance, into a nine-by-eleven-meter main chamber served by a working vertical chimney for a fireplace or kitchen stove, with three smaller secluded side galleries for bedrooms or even a study. Total floor plan: roughly 270 square meters.

In 2008, the remains of one of the former residents surfaced in the cave. Calling them remains is generous, as it was only a tooth and single distal manual phalanx of the fifth digit of a juvenile, or in more normal language, the last joint of a little finger of a child. The layer in which it was found puts its date as being somewhere between 30,000 and 50,000 years old. A single finger bone is enough for paleoanthropologists to classify the former owner as a hominin, a taxonomic division that includes all the Homos, as well as the gorillas and chimpanzees. But for more precision than that, a fingertip is not nearly enough.

Except that it was. The tooth was big, bigger than expected for both *Homo sapiens* and *Homo neanderthalensis*.

But what was in the finger was enough in fact to overturn human evolution yet again. Russian diggers who found the bone passed it on to Svante Pääbo for DNA analysis, and here they were extremely lucky.

Pääbo's team managed to extract the mitochondrial genome from the single tiny bone—in doing so they destroyed it—and the results were published in *Nature* just before Easter in 2010. That sequence, patched up from the inevitable decay of several hundred centuries, was enough to be revolutionary. It wasn't us, and it wasn't Neanderthal. No other species in the genus *Homo* was known to exist at that time in Europe or Asia, and it was not a sequence akin to our primate cousins, chimps, or bonobos. It—she, we would soon discover—was a new type of human.

Of the limited information that was extracted from this little loop of DNA, the number of key differences in the DNA pointed to a human whose ancestors moved out of Africa in a migration that was different from the ancestors of the Neanderthals (around half a million years ago) and our own exodus from the motherland that began at least 100,000 years ago. The number of differences in her DNA was twice as many as between us and the Neanderthals, and that number can be used to calculate the last common ancestor of these three people. Around a million years ago, somewhere in Africa, a group of humans lived who were to be separated into us, the Neanderthals, and the Denisovans.

That separation was temporary, lasting just a few hundreds of thousands of years. By Christmas 2010, the rest of her genome was complete. An "unknown hominin from Southern Siberia" is how she was described in that first study. Fossils are rare by their very nature, and to get fossilized requires a lot of luck. With this finger, one bone out of more than 200 that made up her body, they got extremely lucky. It seems that these conditions preserved the DNA locked inside this tiny bone better than any one of the Neanderthal remains found so far.

The bone size, when compared to Neanderthal or modern human finger bones, indicates that its owner was a juvenile, a child. It says nothing about sex. The DNA says it all. Sex is determined by DNA, but a chromosome count is not an option in ancient samples. Chromosomes only make those nice identifiable shapes at a particular time in the cell's natural cycle, and can only be achieved in cells taken from the living. So a visible Y or a pair of Xs was never in the cards for the Denisovan finger. Instead, Pääbo's team extracted the DNA in fragments and inserted them into a kind of database made of other bits of DNA designed to host the unknown sections. Once this library is established you can use it to generate the sequence of the letters of code on a computer, and make all the comparisons you like without touching the bone ever again. One of the first steps was to look to see if there were any fragments that looked like a modern Y chromosome. The answer was no, so it was very probably a girl. This is an absence of evidence, which we scientists like to remind people is not the same as evidence of absence.

But in this case, the nonappearance of Y is enough for her to be believably female. Other characteristics that emerged from her genome suggest she had brown eyes, brown hair, and dark skin, all traits common to a species of *Homo* before the modern advent of light skin and blue eyes (all discussed in Chapter 2).

The second key comparison was to try to establish what the ancestral distance was between us, them, and Neanderthals. The way to do this is to look at a stretch of DNA and compare the precise sequence in several species. Very simply, the more similar they are, the more closely related the owners are. This applies at every level of living thing, from twins to bacteria. The comparison in the new genomes showed that Denisovans and Neanderthals were more closely related to each other than either was to any living human. But the real kicker came with the revelation that Denisovan DNA was alive and well in contemporary Melanesians—the indigenous people of Fiji, Papua New Guinea, and a scattering of islands off the northeast coast of Australia. Just as the Neanderthals left their permanent mark in me and you if you are of Eurasian descent, these other people, known only from this single bone, imprinted their genetic mark through the ages in the ancestors of these island people, up to 5 percent of their genomes.

The people of Tibet carry adaptations to living at altitude, as they do in the deeply inhospitable plains around Everest with its lower levels of oxygen. The people of China to the north and India to the south do not. Mostly, that adaptation is crystallized in a gene called EPAS1, which sits in a region of DNA whose sequence is notably different from the Tibetans' neighbors. By comparing this highly unusual bit of DNA to other local and global known sequences, it seems that this adaptation was plucked from the Denisovans. Admixture introgression had provided an adaptation that allowed Tibetans to thrive in an environment in which probably you and I would struggle.

We haven't been able to determine how many chromosomes either the Neanderthals or the Denisovans had. It seems likely that they had the same number as us, and interbreeding as we did so successfully reinforces this assumption. There is something significant that we do now know about the large-scale arrangement of the genomes of these peoples. There are four living genera of hominidae, aka the great apes: *Pan* (chimps and bonobos), *Pongo* (orangutans), *Gorilla* (gorillas), and *Homo* (us). The first three have twenty-four pairs of chromosomes, whereas we have twenty-three. But all great apes, including us, share effectively all the same genes in our genomes. The discrepancy is found in our chromosome 2—the second largest single chunk of DNA we carry. It's such a whopper because it is an end-to-end fusion of two chromosomes found in chimps, orangs, and gorillas. We know this because the genes on the two hairy ape chromosomes are effectively the same and in the same order as those found on our chromosome 2. And we can also see the remnants of chromosome architecture, no longer needed after this massive joint. Chromosomes have a pinched waist, like a knot in a long balloon, called the centromere. The two chimp chromosomes have a centromere each, but our fused chromosome 2 has one obvious centromere, and the shadowy remnants of another, betraying our common ancestry with our hairy quadrupedal cousins. We can spot these in the DNA sequence, and there they are in both the Neanderthal and Denisovan genomes. This says that this uniquely *Homo* characteristic occurred before the three of us diverged.

Pope John Paul II pontificated, as popes do, in 1996 that evolution was more "than just a theory," which while being a generous bridging gesture between the magisteria of science and religion, misunderstands that theories in science, unlike in the vernacular, are the top of the intellectual pile, the zenith of descriptions of the true nature of nature. Theories are the best we've got. Anyway, he tried to reconcile the idea of our divine special creation with the irrefutable evidence of our evolution from earlier apes by suggesting that there was an "ontological discontinuity"—if we were looking for a moment where the metaphorical breath of God metaphorically entered us, then it could be when those two chromosomes fused. If that were the case, then the Denisovans and the Neanderthals are in our special club too. Otherwise, there is no discontinuity, just a bumpy slide into our present.

What happened to these peoples? We don't really know. The Denisovans are almost entirely mysterious, and while there are rumors of other bones being found in that Siberian cave, we currently are relying on a fingertip and a big molar. A second tooth was discovered in late 2015, and I am told that more are on their way. We see their genomes, mostly of just that young girl in fact (the first tooth is from a different young adult), and we see that they are neither *Homo sapiens* nor *Homo neanderthalensis* (though the most recent studies show that just like with *Homo sapiens*, the Neanderthals mated with the Denisovans too). They have not been us for a million years, nor Neanderthal for 800,000. Yet, at some point, a group of them hooked up with a tribe of *Homo sapiens* whose descendants ended up populating large chunks of the east, and in particular the Melanesian Islands.

A ghost from our past

There is one more almost surreal twist in this tale, so far (for I am sure there will be many more in the months and years to come). In 2013, David Reich, a Harvard geneticist essential to many of the stories in this book, further looked at the Denisovan genome with a very close squint (in the form of some sophisticated and sensitive statistical analysis) and saw something difficult to explain. The Neanderthals and the Denisovans split off from the lineage that ended in us some 400,000 years ago. But if you look hard enough, the Denisovan genome looks slightly more different from ours than it should. What this implies, according to Reich and others, is that they also had admixed and interbred with another species, one for whom we have no DNA to compare. A ghost population, a group of humans who are only known by the spectral presence they left in DNA as an outcome of sex.

Speculation is rife about who they were. Some, such as the UK's doyen of ancient human bones, Chris Stringer, from the Natural History Museum in London, thinks they might be a species known from bones from which we have not fished out any DNA yet—*Homo heidelbergensis*. There's another possibility, featuring another set of bones that will only further muddy the once-clear picture of human evolution. In the Longlin Cave, in Guangxi Zhuang in southwest China, remains of some odd people have been found. They died somewhere between 14,000 and 11,000 years ago, right on the turn of the Pleistocene to the Holocene. By this stage we see only anatomically modern humans. But these guys were not that. They shared many characteristics with us but also many primitive features. In the cave were the cultural remnants of lots of cooked venison, and they have become known as the Red Deer Cave people. Colin Groves, one of the Australian researchers who catalogued the find, said that they were "close to us but not quite 'us,'" and may be classed as yet another human species. DNA extraction has so far failed, but some have suggested that these people might be the fruit of Denisovan and modern human matings. Until we get the DNA, the answer is another enigma in this increasingly populous field.

So there it is, for now. If you want to stick to the old definitions of a species, then it is impossible to call Neanderthals a separate species. The Denisovans have not yet been classified with the official taxonomy classifications that we adhere to. DNA is not how it is done, and a finger bone and a tooth are just not meaty enough. They were not us, and they were not Neanderthal, but we happily bred with them. This is a strange conundrum that betrays our difficulties with classifying life, and our adherence to a system that was designed to show the perfection of divine creation, organisms static in time and set in stone as they stand before us. Darwin's great idea ruined that ideal, because he recognized that life passes through time, and changes continually. The only life forms that don't change are dead ones.

Obviously, a worm is not the same as a monkey, nor conceivably capable of impregnating one. These are creatures separated by hundreds of millions of years of evolution. We know that chimpanzees and us, only 6 million years apart, cannot interbreed, though we can conceivably have sex, however distasteful that sounds. But what we now know as a result of DNA is that the physical distinctions brought on by tribes of physically distinct human species a million years ago, separating from each other for many thousands of generations and years, was not enough to prevent us from having successful reproductive sex.

The seven billion of us alive today are, according to all the evidence available to us, the last remaining group of human great apes from a set of at least four that existed 50,000 years ago. One of them, the little people of Flores, was remote and unusual, as decreed by the evolutionary weirdness of living an insular life. But the others were not that different from us. In the years to come, we will scrutinize the bones of dozens, or hundreds of ancient humans, remains that have been difficult to place, or controversial, or who have unknowingly caused furious arguments. Some of these battles will be settled with DNA, and others will be stoked. One thing seems inevitable: The ancient genomes left to discover will reveal that the world was a whole lot more cosmopolitan in the millennia before we came to be the last representative of the genus *Homo*. The landscape of the last million years of our past has been flattened by DNA, and the rebuilding has just begun. The family tree is cut down, and we're mapping out the tributaries and rivers and brooks that led to the pool that we all swim in today.

This description is just skimming the surface of our new genetic history. As the techniques become more robust, and bones continue to be unearthed, we will expose more and more sexual dalliances from our ancient past, between all sorts of peoples. The rate at which new studies about the DNA of Neanderthals and Denisovans are being generated is such that on an almost weekly basis, dates, locations, and impact of the relations between those early people are being revised. It's now clear, more than ever before, that the old, simplistic view of how we came to be who we are is just wrong. Gone are the days of neat branching trees, or the hunched ape step by step standing tall. The American beat poet Edward Sanders invented a word that I think is brilliantly apposite here. It was coined to describe a chaotic situation, and is often deployed by the military when maneuvers go badly wrong. For our purposes, if we are to look at the evolution that led to where we are now, instead of the nice neat tree, I think it could reasonably be described as one big, million-year clusterfuck. Whenever humans met—sapiens, Neanderthal, Denisovan—they had sex. What a time to be alive!

We'll discover in the next few chapters that family trees are never trees in the way that we draw them in historical pedigrees, or in science papers when looking at a few closely related species, or even as Darwin scribbled so perceptively in his notebook in 1837. As long as we have been erect, mobile, and frisky, the branches have been entangled within our species, and now we know with cousins that have been apart from us for so long that they looked deceptively different. The Neanderthals, the Denisovans, and our other as yet undiscovered phantom cousins were different from us, and soon we will know them better.* But they also became us, and we will find them in the old bones and inside our own cells. We carry the past with us. There was no beginning, and there are no missing links, just the ebb and flow and ebb again of living through epochs. Those ancient people never went extinct—we just merged.

* The genetics of the dead is a science moving at exhilarating speed, and reports of bones replete with new sources of DNA are being published on a weekly basis. But in April 2017, Viviane Slon, Svante Pääbo, and a Spanish team managed to do something almost magical. They had been to seven caves known to have once been homes to Neanderthals and Denisovans. They took sediment samples and succeeded in extracting wisps of mitochondrial DNA directly from the dirt. There in the cave floors, they found ghostly traces of the genes of woolly rhinos, cave bears, mammoths, and humans. We can now find the DNA of ancient people in places where bones no longer linger, and the door to past lives creaks open even further.

2
The first European union

Loschbour, Luxembourg, 8,000 years before today

Before Europe, before the euro, before the three Reichs—the third and shortest by the Nazis, the second of the German Empire, and the first the 844-year rule of the Holy Roman Empire; before the rats carried plague pathogen hitchhikers who ravaged people wherever they met; before that crookback villain Richard III; before Magna Carta; before the last time invaders conquered Britain in 1066 with an arrow in the king's eye; before the first Holy Roman Emperor, the great European conciliator Charlemagne; before Vikings and northern Scots set foot on the volcanic wilderness of Iceland, the very first to do so; before the Council of Nicaea and the foundations of modern Christianity; before the fall and rise of the Roman Empire, and the great expansive Alexander and his masterful science tutor Aristotle; before the city states of the Greeks; before the Minoans and the Mycenaeans, and to the north the Saxons, Picts, and Goths; before all of these peoples, tribes, customs, cultures, wars, invasions, technology, writing, before all of history, *Homo sapiens* were already here. Europe has been our land, from the eastern Steppes, to the Atlantic tip of Trafalgar, for millennia before the whole of history. Of all the versions of people that emerged in the world, *Homo sapiens* have been the sole human occupants of the continent and the islands of what we now call Europe for more than 30,000 years.

Earlier people had been loitering around the continent for up to 2 million years. These included *Homo erectus*, a species of humans that had great success leaving the motherland of Africa and making it all around the world, as far east as Java, and all over western Europe. One 1.4-million-year-old erectus tooth has been found in a cave in Bulgaria, other remains in Georgia, in France, and dotted all around the mainland of Europe. Evidence of humans in Britain dates back almost a million years, for most of this period connected to what is now the Netherlands by dry land, and they left tools and useful bits of mammoth in the sand and clay of what is now the beaches of north Norfolk. These days, the rapidly eroding coast is exposing these remnants to us. But for all these people and the bits and bobs they left behind around Europe, their bones are too old, maybe too wet, to have preserved their inbuilt genetic historical textbook.

The Neanderthals made their homes all around Europe (see Chapter 1)—in Germany (where the name we gave them comes from), all over France, and scattered in the east of Europe, Wales, Israel, and further toward the east. They were the first true born and bred Europeans: Neanderthals diverged from our best-guess common ancestor, *Homo heidelbergensis* 500,000 to 600,000 years ago, possibly somewhere in middle Europe, and those Germans didn't last later than 200,000 years ago. We now know that anatomically modern humans—that is, us—didn't arrive in their lands until some 60,000 years ago, having slowly moved north and west from the Middle East and before that Africa. We know that we overlapped in time and space with the previous Europeans—the Neanderthals—and we carry their ghosts in our DNA. Coexistence in Europe probably lasted around 5,000 years, which is in evolutionary measure a click of the fingers, but is a time long enough for maybe 200 generations, epic migration, cultural development, plenty of sex and death, and the general business of living. Even though they endure in us, we outlived those people, maybe hunted them, some have suggested even butchered and ate them, and no Neanderthal bones or tools have been found that are more modern than around 30,000 years old.

So, for the time being, the Europe of the last few tens of thousands of years has been our land. In the last decade, and very significantly in 2015, a glut of new research has transformed our understanding of the origins of the Europeans mostly using DNA from ancients and the living, but also encompassing archaeology and language. Between the end of the Neanderthals and the beginning of history, the settlers in Europe changed physically and culturally, and over the next few pages, we'll see how genes and culture, primarily farming, shaped modern Europe. Just as with our very ancient ancestors in Chapter 1, we're reconstructing the past with old bones and DNA combined, but the bodies are few and far between. "Far between" is less of an issue, as distance between bodies can give us a good idea of distribution over a continent. "Few" is more of a problem, because genetics is a comparative business, and the more data we have, the better resolution the picture is. But that's the reality of people who've been corpses for thousands of years.

The oldest genome of a European comes from a 37,000-year-old square-jawed man who washed up on the banks of the mighty River Don in southern Russia. He's called Kostenki today, and his DNA showed similarities with more recent European hunter-gatherers, as far afield as in Spain 30,000 years later, but few with East Asians. We know that key physical characteristics synonymous with eastern Asians arose some 30,000 years ago. Thick straight hair, a density of sweat glands, and a particular tooth shape largely specific to the people of the east appeared in

China at that time, though these visible changes all stem from one single gene, and the overall genetic similarities and differences between peoples are measured at a far greater depth than skin deep.

The record for the oldest complete modern human genome comes earlier and from the banks of another mighty Russian river further east. The Irtysh is the chief tributary of the Ob, which together cut thousands of miles across the middle of Russia from China to the Arctic Kara Sea. A Russian artist named Nikolai Persitov carves jewelry and sculptures out of mammoth bones, and while on the hunt for these raw materials in 2008, stumbled over a femur—a thighbone—jutting out of the eroding banks of the Irtysh in a southern district. It was of a 45,000-year dead man, now known as Ust'-Ishim after the place of his resurrection. He gave up his genome to Svante Pääbo and David Reich, and their teams in Leipzig and Harvard, who together assembled his complete DNA as we would do with a living person. Ust'-Ishim's DNA shares similarities with both eastern Asians and western Europeans, unlike Kostenki, who had little genetic resemblance to the east. Ust'-Ishim is therefore likely to be ancestral to both Europeans and Asians, and so represents—at least genetically—the people who had migrated up from Africa before one group, over thousands of years, turned toward the dawn, while the other one went into the sunset.

The majority of the latest round of genetic analysis of Europe has focused on the last 10,000 years, a time from which the evidence of farming around the world is strong. This includes domestication of various animals, swine, goats, sheep, llamas (dogs had come earlier, maybe 15,000 years ago, as companions for protection and hunting, fed with the scraps of the spoils), and the cultivation of plants, and processing them with Neolithic tools. Why we began farming is disputed. The climate was more stable during this time in Europe, possibly wetter, and some of the big animals that flourished during the cold—woolly rhinos and mammoths, and other megafauna—were all set for extinction. Domestication might have been easier and needed as Europe became more stable and temperate. There may have been health benefits for a regular food chain, but this has been disputed as some archaeological remains suggest that people got shorter after domestication, which can be associated with a less nutritious diet. Please ignore this sentence, as it will only be of relevance on page 77. The "Paleo Diet" is a popular fad that eschews processed foods and carbohydrates in favor of the only foods imagined to be available to the hunter-gatherers of the Paleolithic: no dairy or processed grains, no lentils, beans, peas, or other human-designed veg. Nuts are OK, but no peanuts, as they're a farmed product. It is almost certainly built on bunkum foundations, as indeed most fad diets are.

By the time of the agricultural revolution, we see multiplication and expansion of genes that encode salivary amylase, an enzyme in your spittle that initiates the digestion of complex molecules. Some people have eighteen copies of it, but chimpanzees only have two. Amylase digests starchy, carbohydrate-rich foods, and helps generate glucose from them, which would provide much needed energy for the evolving and highly energetic brain. It works many times better on carbohydrates that have been cooked. We're not sure when we began cooking; the range of evidence is wide, but food cooked on fires was definitely part of our menu by 300,000 years ago, so before we were anatomically modern. The expansions of these amylase genes, the presence of cooked food, and the massive expansion of our brains that is observed over our Paleolithic evolution point to a positive selection for eating nutrient-rich tubers. This is hypothetical at the moment; it fits the picture quite neatly, and fits into our overall picture of how culture affects genes and vice versa. Diet is not easy to validate in the deep past, but the clues from genetics suggest that the basis for the Paleo Diet is nuts, even if you're not allowed to eat peanuts.

As with all fad diets it probably works a bit, but not because of the content of the diet itself, but because the act of dieting prompts people to eat less and think more about their food, and not to shovel huge portions of pasta or chips on their supper plates. So go ahead and diet, but don't pretend that it's based on some evolutionary precedent. And remember that whatever we did in the deep past, we live longer and better now than at any point in the history of humankind.

Regardless, the agricultural revolution that occurred at the beginning of the current epoch, the Holocene, coincided with the first evidence for farming, even though the reasons for this revolution remain unclear. But it did irreversibly change everything. This transition to a domestic life fundamentally changed us in our bones and our genes, as we'll see soon. The land changed more obviously, as you would expect, when it became worked in this most unnatural way. I like to travel on trains in Britain, and marvel at the green rolling hills. I pine for them when I'm abroad. I like to ponder how entirely unnatural they are, how they've been designed and built over thousands of years, how the hedgerows—so critical to the biodiversity of the land—were put there by people to separate crops and animals and predators and property. Even the coarse wild brush in the highlands of Scotland and much of northern England has been grazed and unnaturally tended and grazed again continually for millennia.

Estimates are that the hunter-gatherers who were all but wiped out by the agricultural revolution numbered around 2 million 12,000 years ago. Agriculture spread like a virus over the continents from its birth somewhere in the Middle East (and dotted in other spots in Africa and China), and would be the dominant business of humans for most of the rest of history. Farming today is industrial, and dominated by monolithic corporations who control almost all the food we eat. But this book is not about that. In its inception, farms were subsistence smallholdings, territories carved out of cleared forests and walled fields, and permanent residences. Crops, by their nature, are seasonal, and with farming came a need to plan, to store foods in pots and jars and silos for lean years. With these plans came surplus, some years, and that draw would bring others into the bountiful communities, which would grow and flourish. We humans are a technological species, science is in our souls, and farming is a technology quite the opposite of natural. From humble beginnings, innovation in tilling the land radically brings greater harvest efficiency. With that comes economic disparity: Some have surplus, others have less, which means some families get bigger, and have more children who live, more culture, more technology; and the cycle continues. By the time of the Roman Empire, there were a quarter of a billion farmers on Earth, and foraging was a living now limited to a couple of million people at most, flung far into the corners a long way from Europe, in Australia, South America, and in pockets in Africa.

Argument about the impact, locations, and breadth of the agricultural revolution has preoccupied anthropologists for decades, and is set to continue. We can now add DNA to our arsenal of evidence, alongside—not in competition with—the archaeology of ancient farming, the lie of the land, the broken pots and pans of kitchens and larders, and of the lovely bones, both ours and of the beasts we kept. Arguing about agriculture is a European preoccupation. For forty years we've been bickering about farming in the European Union, and the Common Agricultural Policy.

This tale is also all about farming policies, and it starts not far from Brussels. Eight thousand years ago there were probably about 5 million people on Earth, the current population of Norway. They were spread far and wide. By this time, people had reached the tip of South America, Australia, and most places in between (though not into the South Pacific islands or New Zealand), and we have hundreds of corpses and tons of archaeology from these times. We've only just begun harvesting their genomes, yet already the picture of the modern evolution of us is being rewritten. Much of the work has been done in large collaborative groups of geneticists, archaeologists, and historians, and the powerhouse of the genetics work is headquartered in David Reich's lab in Harvard. Reich's international team plucked the DNA from nine people in 2015. One was a man from a murky cave dwelling in a village outside Loschbour in Luxembourg. One was a woman from Stuttgart. The rest were from a cave in the small Swedish town of Motala. They've all been dead for more than seventy-five centuries, but, though they were almost certainly unaware of it, they were standing on the edge of a revolution that had already begun, and would shape Europe and the world for evermore.

Loschbour was a hunter. In his final resting place, he was surrounded by the paraphernalia of a life stalking and killing animals, flint blades knapped to mount on spears and be used to butcher the meat and tailor the hides. In the flatlands of northern Europe 8,000 years ago, these beasts would be wild boar and deer. He was one of the last, though, as immigrants were coming from the east, with new customs, new knowledge, and a penchant for herding rather than hunting.

Loschbour was pulled from his grave in 1935, but DNA was extracted from an upper molar in 2014. Stuttgart emerged from the tomb in 1982, and had her DNA taken from a molar too. She was a farmer from around 5,000 years ago, and was found with the paraphernalia typical of the Linear Pottery Culture—decorated pots and gourds, stone tools, and evidence of animal husbandry. The Motala clan were disturbed from their mausoleum only in the last decade, and samples were taken from the teeth of nine skulls and the tibia and femur—the shin and the thigh bone—of an individual. Like Loschbour, they were hunters too.

All three were genetically different. Stuttgart and Loschbour had genes for dark hair, Loschbour and Motala had genes for blue eyes. Stuttgart had sixteen copies of the amylase gene indicating a carb-rich diet; Loschbour and Motala had fewer. These indicators paint a crude picture of what they looked like, and a more detailed image of some of their cultural practices. But by interrogating the subtleties of their DNA, David Reich and his colleagues were able to unearth the foundations for 10,000 years of European occupation. Our understanding of the genetic basis of Europeans had come from studying the genes of modern, living Europeans. This has plenty of merit, but we now know that the DNA of a population today does not necessarily reflect the DNA of the population in the same spot from thousands of years ago. By comparing the genomes of Loschbour, Motala, and Stuttgart with 2,345 modern Europeans, Reich could pick out the various contributions that these early Europeans have made to the overall structure of Europeans today. What they showed was that we Europeans are drawn from three different groups of people. They're not the ones represented by Loschbour, Motala, and Stuttgart, but the differences between them highlight different proportions of DNA that we see today.

The first European *Homo sapiens* were the hunter-gatherers who had moved up from Africa via central Eurasia 40,000 years before, and overlapped and mated with the indigenous Neanderthals. Between 9,000 and 7,000 years ago, we see the genetic fingers of eastern farmers reaching into this population. They didn't usurp them or wipe them out, though. We see the two populations living not exactly side by side, but at the same time, some hunters, some farmers, and we see the slow integration of genes from the hunters enter into the genomes of the farmers.

And then around 5,000 years ago, another major wave of easterners arrive. The Yamnaya came from the Russian Steppes, driving sheep, riding wagons, making bronze jewelry, and covering their dead in ochre as part of ritual burials. They came and rapidly their way of life spread into middle Europe, bringing their culture and genes, and their fair skin. Farming came to dominate and eventually entirely replace hunting and gathering. Pale skin came to replace dark skin, and we will see more of that later in this chapter.

If you are pale skinned, you are almost certainly a product of these three waves of European immigration. We are in a process of reevaluating human history through these new models of gene flow. There was a time, not that long ago, when we assumed that the history of humankind was simply a series of what evolutionary biologists call "founder events"—small tribes moving further from the motherland of Africa, setting up camp, growing, and then budding off to form new small tribes who would repeat the process in new uncharted territory. Early genetic studies (meaning "in the last decade," such is the ridiculous pace of this science) indicated that this model may well have been accurate, as small samples of chunks of repetitive DNA seemed to cluster together. You would also expect to see a decline in genetic variation the further we moved from Africa, which is broadly what we see. The latest analyses incorporate the fact that current residents of a geographical area are not necessarily very good representatives of the residents of the deep past. This is obvious if we look at areas that have been subject to migration by modern Europeans: The majority of the peoples of Australia or North America today are from Europe in the last 500 years, and so their genomes are not representative of the indigenous people who were there first. But this impermanence is also not necessarily true for much older populations of humans. The assumption, for example, that the Siberian farmers are going to be most similar to the first settlers of the American continent because of their proximity to the Bering land bridge 15,000 years ago (now the Bering Straits) is not correct. In fact, digging up old bones and wheedling out their DNA shows that today's Siberians are more like East Asians, but the ancient Siberians were more like Native Americans, mixed in with some northern Eurasian. All this means is that we made assumptions about patterns of migration that were much more linear and spread like ripples, rather than the picture that has emerged in the last couple of years, which says that we moved in all directions all the time, and laid our hats and flowed our genes in a matted crisscross, instead of a nice clean radiation. Oceans and mountains are good barriers to gene flow, but on big open continents the horizon is the limit. Genes flowed out of Asia and into the Americas.

These ancient genomes don't just tell us about movement and migration. DNA is one of a suite of tools we use now for reconstructing the past, and we build up these pictures with the agglomerated knowledge from archaeology and geology. DNA also reveals behavior. Culture can become embedded in our cells just as it gets buried in the floors of caves, bogs, and dwellings.

Milk (and honey)

It is a strange thing that that some adult humans drink milk. Mammals are, of course, defined by lactation, by mammary glands and the suckling of our infants. It is specifically for neonates, and when mammals grow up, they put away this childish thing and stop drinking milk.

But not everybody does. In the West it's fairly normal to drink milk in various forms into adulthood. Europeans and their American descendants do, and some African and Middle Eastern pastoralists do it too. But it was not always thus. Most adult humans today, and almost every human in history, do not even have the capability to digest milk. We all have an enzyme called lactase, encoded by a gene called LCT, and its sole job is to digest milk. The sweetness in milk comes from a sugar called lactose, and lactase seeps out of your stomach lining and slices lactose in half to produce the sugars glucose and galactose. Elegant names are not always a preoccupation of biologists.

For most of human history lactase has been active only in babies. After weaning, the gene's activity is radically reduced, and as a result, for most adults, for most of human history, milk has been off the menu. Most people, for most of human history, will have experienced a full deck of problems that come with drinking milk past weaning. Symptoms include bloating and cramps, vomiting, diarrhea, flatulence, and borborygmus, which is a technical word for a rumbly tummy. The absence of lactase, or its reduced activity, means that the lactose doesn't get digested in the small intestine, so it passes into the colon, where it encounters bacteria that can break it down and it ferments, causing gas buildup. That's the direct cause of the bloating and fartiness, but also the increased pressure triggers diarrhea, and so on. This is called lactose intolerance, and admittedly though not particularly pleasant, it's not the worst condition someone can have, and is pretty normal for most people if they drink milk into adulthood. Which is why most people don't.

Except if you're of European descent. Your lactase continues to work throughout your life. This unusual phenomenon is called lactase persistence, and although a splash of milk in tea is the English way, and even a mug of hot chocolate might seem very normal to us, we are the weird ones. There are a handful of African populations, some in Southeast Asia, and a few Middle Eastern peoples whose lactase persists, but for the majority of modern humans, milk equals tummy troubles.

The genetics of lactase persistence are well understood. A very small number of individual changes to the DNA in and around the lactase gene account for its persistence. The vast majority of Brits, and northern and western Europeans (including places colonized by them) have a single change, a C becomes a T, around 13,000 letters of DNA before the start of the lactase gene. This type of mutation is not uncommon, but it's really where the language metaphor to explain genetics falls down. Although one hopes that the coherence of the chapters and paragraphs and sentences in this book has an overarching theme, in general each idea follows on from the last. In genetics, though, one gene can have a direct effect on another, yet can be miles away on the same chromosome, or even further away on another chromosome. So the random sentence 13,000 characters ago on page 69 was bizarrely irrelevant and totally ectopic then, and I instructed you to ignore it. But it is directly related to this one, and that is precisely what happens frequently in the genome.

Thirteen thousand nucleotides upstream of the beginning of the lactase gene is a region that controls its activity, and a mutation in that distant control center accounts for the vast majority of milk drinkers. By looking in this region for this and other minor variations that have no real effect, and comparing these accumulated changes with other genomes around the world, we can make estimates of how old the lactase mutation is, which comes out in the region of 5,000 to 10,000 years BCE. And the presence of these particular genotypes altogether in a cluster that includes the lactase gene and other bits and genetic bobs, indicates that it was all seriously favored by natural selection.

Now, unless it conferred some other advantage (for which there isn't any real evidence), it's hard to see what the evolutionary advantage might be for lactase persistence in the absence of a regular supply of fresh milk. And so we think of this as a classic example of how we have invoked shifts in our genome with our own practices—a gene-culture coevolution—experienced only in communities that were practicing dairy farming with domesticated milky beasts. What advantage having both access to milk and the ability to process it might seem obvious: In fact, it's really the realm of intelligent but speculative guesswork. A regular supply of nutritionally rich food is one; avoiding the boom and bust cycles of seasonal crops is another possibility.

By 6,000 years ago, milk had become a part of Neolithic life. Shards of crockery recovered from digs in Romania, Turkey, and Hungary hold traces of caked gunk, and these got squirted into a gas chromatograph by Richard Evershed's team at the University of Bristol. In that long, extremely thin tube, carried by an inert gas such as helium, the ingredients of these residues are separated into their constituent molecules, which move at different speeds as they trickle away from the carrier, and with a reference guide, can be precisely identified. In among the detritus of being buried for several thousand years, there was the signature dairy fats. There's not many ways they could've got there unless they were used to store milk. By 5500 BCE, we were making cheese. Sieves and pottery colanders resembling modern cheese strainers had been found in Poland, and in 2012, again, telltale residues were scraped off these ancient dishes. The suboptimal washing-up skills of the people who owned this crockery again revealed fat from milk. Cheese, of course, is a strange thing in itself, and odd that we should eat it. It's milk that has gone bad, probably the first processed food, but it may have been a useful way of storing the nutrient-rich milk in solid form, possibly more like a glob of mozzarella than a wheel of Stilton. We know that, much later, the Romans ate goat's and sheep's cheese, but used cattle primarily as beasts of burden. They noted that people of Germany and Britain liked to drink milk.

These examples of dirty dishes left for millennia are few and far between, and we need other ways to peer into the past. Lactase persistence is now effectively universal in Europeans (those African and Middle Eastern peoples who also are lactase persistent have different mutations and are unlikely to have a common origin). At some point, maybe 8,000 years ago, it was absent. So the question is to pin down when it emerged and how it spread. In 2007, Anglo-German geneticists dug up bodies from eight locations in eastern Germany, Hungary, Lithuania, and Poland, all from between 5,800 and 5,000 years ago. The Germans' graves are from cultures of people who were potters, and are known to have had farming practices, probably the oldest in northern Europe. From their teeth and ribs and other bits of bone, the scientists extracted ancient DNA. None of these guys had the mutation that endows the owner with the ability to process milk.

These points of data, genomic and archaeological, set up the rough time frame. They are, admittedly, somewhat imprecise, but such is the nature of the past, truly a foreign country. Where dairy farming began would be more difficult to ascertain.

But not impossible. Mark Thomas at University College London has been in pursuit of the origins of dairy for years (alongside his many other ancient genetic detective works). In 2009, he put together the pieces of genetic and archaeological evidence into a computer model, a kind of statistical jigsaw puzzle, from which the clearest picture of the roots of milk farming would emerge. The distribution of lactase persistence genes in and around northwest Europe today, including Scandinavia and Ireland, had suggested that its origin might have been around there. There had also been suggestions that its evolutionary advantage might have been to do with lower levels of vitamin D in these northern tribes; milk drinking can compensate for that obscuring of the sun.

But that's not what Mark Thomas' team found. They computer simulated a scenario of how lactase persistence might have become so widespread by plugging in archaeological data—such as the carbon 14–dated presence of the artifacts of dairy farming, with a careful comparison of the specific genetic differences between three groups of ancient people—foragers, non-dairy farmers, and dairy farmers. We know that the lactase persistence gene was selected by evolution, and didn't just drift into ubiquity, so the simulation trots out a setup where that version of the lactase gene supersedes the versions in the hunters and milk-less farmers. It maps not where it arose, but where it began to fill the population at the expense of earlier types. When collected, the simulations plot a map showing the highest chance of the evolution of the lactase persistence allele, and it locates it in an area engulfing Slovakia, with Poland to the north and Hungary to the south. This fits with the archaeology, and the residues found in those Hungarian and Polish farmyard digs. At 7,500 years ago, these people were farmers with structured garden farms, where they grew wheat, peas, lentils, and millet. They husbanded cattle, swine, and goats, and occasionally hunted boar and deer on top of their agrarian lifestyle. They used flint and wooden tools, but not metal, and used earthenware vases, jugs, and pots with lined designs, from which we derive their name: Linear Pottery people.

The same Bristol team that scraped the gunk from the dairy pots also inspected ancient crockery for other clues to the diets of our ancestors, and in 2015 found the taste of honey. They analyzed thousands of fragments of pots and pulled off traces of beeswax from just four, dotted as far afield as Turkey, Denmark, and Algeria. In other studies they found the remnants of meat. This collected diet data shows a picture of a people cooking and farming, collecting honey, herding beasts—boar and cattle—and milking them, storing that milk in pouches and pots, sieving the curd from the whey. And as those skills developed and spread across the continent, so did the gene that renders that dairy produce so valuable to us.

There are two important lessons in this story. The first is the answer to a question that people often ask: Are we still evolving? It's hard to determine in the present, because we can't very easily see the changing flow of different versions of genes in populations around the world. Evolution is a fourth-dimensional process, it occurs over time, as well as in space, and it happens slowly in most cases. So we have to be patient in assessing its impact, and at the moment we simply do not have enough data. But we can definitively say that we have evolved in our recent past, during a time when we were who we are today. The pressures of natural selection have undoubtedly changed as a result of the evolution of our behavior and culture, but milk shows that, just as in every organism that has ever lived, over time our genes continue to change.

The second point is that the world in which we live is shaped by us, by our practices and culture, by our very existence, and our DNA responds to that in turn. Genes change culture, culture changes genes. Farming has been arguably the single greatest force in changing human culture and biology. So, just think tomorrow morning when you're pouring milk and dribbling honey into your porridge, that it's your ancient European ancestors (if indeed you are European) who began a revolution in culture and in your DNA that allows you to tuck into your breakfast. *Bon appétit!*

Blue-eyed blonds

One of the more obvious differences between modern humans is skin color. We make crude visual distinctions and effectively meaningless categorizations based on average skin tones, such as black or white. The question of race is explored in depth in Chapter 5, and I will explain why geneticists ascribe no scientific value to these broad racial attempts at definitions of peoples. In parts of Africa and Australia and the islands of the Indian Ocean, people have the darkest of dark skin, almost black in its true sense. In Scotland, or the north of Sweden, some are so pale as to be almost translucent. All skin tones in between exist in abundance. This is not some liberal fantasy, it's simply a truth that a Pantone swatch of human skin is a continuum.

Nevertheless the physical characteristics of the Europeans are unusual, just as their love of milk is. We can broadly say that many Europeans are blond and some have blue eyes with pale skin. Some are even redheaded. These are mutations in the genetic sense; as mentioned before, this simply means the DNA—the genotype—is altered from what came earlier, and the physical result of that change—the phenotype—is different.

At least eleven genes appear to have a direct role in determining skin and hair color. Overall, it's all down to the density and type of melanin that you produce. There's two types of melanin: eumelanin (pronounced you-melanin), which comes in either black or brown, and phaeomelanin, which is reddish and is the root of ginger hair, which we'll come to shortly. Specialist cells called melanocytes sit at the base of the skin, and produce melanin following exposure to ultraviolet light from the sun. The melanin collects in little specialized pockets called melanosomes, which shuffle between cells, which is why people tan in the sun. They move into position atop the cell nucleus as a means of protecting the DNA within from UV damage, which can slice through the double helix with abandon. In hair follicles, the melanocytes transfer the melanosomes into the base of the growing hair, and that is how your hair is colored.

The melanin is the end point of this sequence of metabolism. The range of hair and skin colors is partly dependent on the types and density of melanin you have, but also a whole battery of other genes that seem to influence how your melanin plays out in your hair and skin. The protection that eumelanin offers to the cell nucleus is an adaptation to sunny weather. Pale skins tan to protect themselves, in a somewhat post-horse-bolting way (so my advice is wear sunscreen), but those in the blazing sun near the waist of the earth are preemptively protected by being full of melanin.

With the genes identified, we can, with relative ease, say when these changes happened and where. Of course, it's not at all straightforward, and they appear at different times and in different forms. But we can say with confidence that the Africans who populated the Middle East and southern Europe 50,000 years ago were dark skinned. We also know from the DNA of burials in Hungary and Spain and the man from Loschbour in Luxembourg that these hunter-gatherers had dark skin around 8,000 years ago. There's no trace of genetic fair skin between those two dates. In 2015, Iain Mathieson (and a host of others) identified two important genes with tedious names, SLC24A5 and SLC45A2, and showed they also did not have variants that are associated with depigmentation, that is, pale skin. Seven thousand seven hundred years ago, the Swedish clan in the Motala cave had both pale skin versions, as well as a version of the gene HERC2/OCA2 that also plays a role in light skin and blond hair, but also clear blue eyes. The Swedes have looked like that for a long time.

The new genetics created by genomics allows us to move from speculation about characteristics that might be under the pressure of evolutionary selection to certainty. The most comprehensive study to date arrived in winter 2015, and brought with it the largest European resurrection on record. Iain Mathieson's landmark paper contained the genomes of 230 people who had died between 6500 BCE and 300 BCE. Some of them we knew about already: Loschbour, Motala, and some of the Hungarian and Polish bodies were in there, but dozens of newly discovered, long-time dead were invited to the party, at least the bits of DNA the scientists could finagle out of the densest bone in their skulls. Reach behind the fleshy lobe of your ear, and you'll feel a hard bone rising from your skull. This is called the mastoid process. About an inch directly inside your skull from there is a hard ridge of bone called the petrous. Its name comes from the Latin word *petrus*, meaning rock, as in petrified and Peter, the rock on which the Christian church is built. This is because it's rock hard. It is therefore a treat for preserving DNA.

So, in this mix were Anatolians, Iberians, Yamnayans, Poltavkans, and Srubnayans—all peoples from the coast of northern Spain to the Altai Mountains in Siberia, some 2,300 years old, some 8,500. With a gang this big, the signs of natural selection become easier to detect, and that's what they fished for. Again indicative of how far this field has come, Mathieson et al. hunted for patterns of natural selection among more than a million individual changes in the letters of DNA that we call SNPs across all these genomes. Lactase persistence came out top, and a few other diet-related genotypes, almost certainly adaptations to changing diets brought on by agriculture. A couple of them are variants most closely associated with vitamin D, again suggesting a change in diet, and in our changing ability to acquire essential dietary supplements.

They saw selection for DNA associated with celiac disease, a form of irritable bowel sensitivity to the wheat protein gluten, but it's likely that this was carried along by natural selection with other dietary adaptations; one variant found helps process an amino acid called ergothioneine, which is not abundant in wheat. Early farmers of cereal crops may have found advantage in this allele, and an irritable bowel came along for the ride. They saw the selection of genes for lighter skin and blue eyes, though we know that the people of Motala were already blue-eyed blonds. Light skin from the east spread quickly through Europe, and so the population structure over time was upheaved with a change in people, how they looked, and in agricultural practices. They looked at height too, which we know is a trait heavily influenced by genes, and a little but significant amount by the environment (see Chapter 6). The Yamnaya were nomads and pastoralists, probably shepherds from the western steppes of Russia, who we know were tall from their bones. We can also spot tall genes in their DNA, which they brought with them from the east 4,800 years ago, and these spread alongside all the farming and dietary changes that were sweeping through the continent. That was mostly in central Europe and to the north though. In Italy and Spain, selection favored the short, possibly because of colder weather and poor diets.

The picture of 10,000 years of European union slowly comes into focus. More than anything specific, what these studies show is evolution by natural selection in action in the last 10,000 years. It shows the interplay between movement of people, introgression of cultures, particularly farming cultures, and their profound, measurable effect on the most comprehensive record we can muster. In the genomes of the dead we can see natural selection at work.

We can't necessarily see why particular genes are being selected though. These are not the types of adaptation that we might glean from natural history programs on television—big visible changes like tusks for showing off to females, or a variety of beaks for highly specialized nut-cracking as you might see in Darwin's famous finches. But they are Darwinian evolution nevertheless—subtle changes in DNA, and subtle changes in how common that DNA is in breeding populations. The spreading of alleles through a population is the measure of natural selection. There will be many more signals of selection as we unearth more bodies and get more DNA from our European ancestors. These techniques will be and are being applied across Asia and in the Americas, and into the last places on Earth that humans migrated to, the South Pacific and

New Zealand. The mix of history, archaeology, and now DNA is building a new picture not just of migration, but of the evolution of us—how we came to be what we are.

The reds are coming

"People who haven't red hair don't know what trouble is."
L. M. Montgomery, *Anne of Green Gables*, 1908

Extinction should be considered troublesome, so Anne was right, if the newspapers are to be believed. Pale skin and blonds may be an adaptation to the north, and compared to the hair of Africans, eastern Asians, southern Asians, and the indigenous people of the Americas, flaxen hair is a rarity. But there's an even more unusual hair, and that's red. Red hair is caused by changes in a single gene, and exists in the overall global population at about 4 to 5 percent, making it beautifully unusual. Its increased prevalence in Scots (and the Welsh and English, and other northern European populations) is probably due to a degree of isolation in an ancestral group at some point in our ancient history, but we don't really know. Around 40 percent of Scots carry at least one copy of this allele, and one in ten are redheads, but worldwide it is the most unusual hair color.

There are interesting stories within this gene. The protein it encodes is called melanocortin 1 receptor (MC1R), and belongs to a broad class with the equally unwieldy name G protein-coupled receptor. These are long, bendy molecules that straddle the cell membrane, and upon receiving the appropriate molecular signal from outside the cell, trigger a metabolic pathway. In the case of MC1R, a molecule sent from the pituitary gland to melanocytes prompts these cells to produce melanin in skin melanosomes. Though most people on Earth produce eumelanin, which is brown or black, in people for whom their MC1R contains a redhead mutation, phaeomelanin is produced. The melanosomes feed into the base of a hair follicle and this is what makes redheads redheads.

Of course, as is always the case in human genetics, it's not quite as simple as this, and much more interesting. The protein is 317 amino acids long, and there are several different mutations, all of which switch eumelanin to phaeomelanin. All human proteins are made up from different combinations of 20 amino acids, each of which is encoded in three letters of DNA in a gene. In MC1R, if at position 151 you have the amino acid cysteine instead of the more common arginine, you have red hair. If at position 294 you have a histidine instead of an aspartic acid, you have red hair. There are several other mutations that I won't list here that have the same effect, but this goes some way to explain why not all red hair is the same.

In June 1997, J. K. Rowling introduced the world to a boy named Harry Potter and his best male friend Ron Weasley. Ron has many ginger-haired siblings, including identical twin brothers Fred and George. Three weeks after that first Harry Potter book was published, and presumably a magical coincidence, the first major study of ginger-haired twins was published, including twenty-five pairs of Weasley-ish twins. Three major variants were identified that associated very strongly with red hair, but something interesting emerged from the control group. These included dizygotic twins who were discordant for red—that is, nonidentical twins, one a ginger, the other not. Five out of thirteen of these tested had identical MC1R genes. As was beginning to become abundantly clear at that time, just the presence of a redhead allele was not enough to guarantee a redhead. Genes never work in isolation, and almost never have just one role. The discordance of these twins showed that there must be other genetic modifiers that powerfully influence the expression of the MC1R gene to the extent that the phenotype can be either red or not.

Samples taken from a couple of Neanderthal genomes (one from El Sidrón in northwest Spain, another from Riparo Mezzena cave in Italy) indicate that their MC1R had an alteration at position 307 (a glycine where we have an arginine). As mentioned in Chapter 1, this variation is not found in modern humans with red hair, and there is no Neanderthal hair that has survived the ravages of epochs. But there are some cunning tests we can do to try to work out the color scheme. By inserting the version of the protein that these Neanderthal people had in Spain into cells in a petri dish, we can see not the color itself, but the activity of the cells and then speculate about the color that might result from the cell's behavior. The different mutations we see in living redheads can reduce the function of melanocytes in different ways, and indeed these cellular tests show a reduced function of the melanocytes. But does that mean ginger? Possibly. The physical distance between the source of the two Neanderthal genomes sampled suggests that they weren't a couple of freaks, and that we just happened to sample unusual DNA by chance. While blond and pale skin is almost certainly an adaptation to northern exposure, the variant we see in these two chaps is unlikely to be. Think of the tar-black hair of most Italians and Spaniards today. We piece together the past with the clues we can find, and build up a hypothesis we can test and puzzle over. In this case, the truth is that we don't really know—for now.

In July 2014, the world woke to the shocking news that the perfect storms of climate change and genetics had conspired to mark ginger hair for extinction. The first headline I saw was from the Scottish paper the *Daily Record*, with understandable concern given the prevalence of ginger in Caledonia. A scan revealed that every mainstream British newspaper carried the story, the *Daily Mail*, the *Times*, *Guardian*, *Telegraph*, *Independent*, *Mirror*, and the *Sun*, with various pictures of sexy redheaded celebrities, frequently the actors

Christina Hendricks, Julianne Moore, and Damian Lewis. Or Prince Harry. Social media and news websites were ablaze with horror. Around the world, *National Geographic*, the *Week* and a host of other apparently sensible magazines and news outlets reprinted the story. The headline in the *Independent* was typical:

GINGERS FACE EXTINCTION DUE TO
CLIMATE CHANGE, SCIENTISTS WARN

Broadly, the newspapers were reporting that, according to researchers, climate change is going to make Scotland less cloudy and more sunny. Therefore, the selective pressure that nurtured the allele for red hair is eliminated, and red hair will no longer be of any use to bearers, and will drift off into the great evolutionary dustbin of once-useful traits. This is an excerpt from the article in the *Independent*, followed by a gallery of famous redheads:

> Dr Alistair Moffat, managing director of Galashiels-based ScotlandsDNA, said: "We think red hair in Scotland, Ireland and in the North of England is an adaption to the climate.
>
> "I think the reason for light skin and red hair is that we do not get enough sun and we have to get all the Vitamin D we can.
>
> "If the climate is changing and it is to become more cloudy or less cloudy then this will affect the gene. If it was to get less cloudy and there was more sun, then yes, there would be fewer people carrying the gene."

Well, no.

Alarm bells started to trip for many familiar with the themes of genes and natural selection. As mentioned earlier, there is good evidence for pale skin being an adaptation to the cold of the north, but none that red hair is too. Furthermore, I'm unaware of any evidence that says that Scotland will become less cloudy as a result of climate change.

Who is Alistair Moffat, and what is ScotlandsDNA? In fact, it is a genetic ancestry testing business, a partner company to BritainsDNA (see Chapter 4), and Alistair Moffat is their founder and chief executive. The story was drawn from a press release from ScotlandsDNA, which coincided with the promotion of a new additional service that tests for the presence of red-hair alleles in a customer's genome.

Alas, a fiction can fly around the world before the truth has managed to pick the sleep from its eyes in the morning. Many condemned the errors inherent in the content quickly, and focused on the discomfit of PR dressed up as research that journalists sometimes fall for.

In this particular case, it is hard to think of a way that the ginger allele might be extinguished. The selective pressure asserted in this tale is that redheads exist as an adaptation to cloudy weather in Scotland. There is no evidence for that. In order for these pressures, if we imagine they are real, to push the ginger allele toward extinction, ginger hair would have to be a powerfully maladaptive condition, meaning that it causes harm to the individual in the environment in which it exists.

I need not point out that being redheaded is not a maladaptive condition. It's a very lovely condition. It is an absurdity, offensive to both redheads and geneticists—a group that contains both family and friends—to suggest that red hair might be subject to a force of natural selection so powerful that oblivion awaits. Even actually maladaptive genetic traits, actual diseases with well-understood modes of inheritance, such as cystic fibrosis or Duchenne muscular dystrophy are not likely to go extinct, because carriers of a single copy live healthily and pass the faulty gene on to their children. Via genetic screening and expert advice, these conditions will diminish, we hope, but the prospect of full eradication is as yet unlikely.

After careful consideration lasting no longer than forty-six seconds, I thought of three plausible ways that redheads will vanish from existence:

1. Humankind goes extinct.
2. Not only would every redhead have to permanently stop having sex for some reason, but every carrier of the gene would too* (which means that everyone on Earth would need to be tested).
3. Redheads, and all carriers of a red allele are exterminated (which also means that everyone on Earth would need to be tested).

To be frank, option 1 is the most likely. The absolute truth is that, like so many human characteristics, we don't genuinely know if ginger-hairedness is a mutation that has a physiological advantage. It is unknown if it is an adaptation to northern climates, to the beloved gray weather of Scotland or Scandinavia. It might be sexually selected, as despite the lazy mockery that red hair sometimes invokes, some of us find it enormously attractive. It might be that the mutation, random as they all are, was neutral, and had no noticeable effect, but drifted in populations largely isolated somewhere in the northern realms of Europe, and became fixed in these populations at the low, special frequencies we see today.

Such is my confidence that red hair will not go extinct as predicted by ScotlandsDNA, I publicly offered, in an article in the

* In the episode "Ginger Kids" of the always-brilliant cartoon *South Park*, one of the lead protagonists, Eric Cartman, who is possibly the most evil creation in the history of fiction, is tricked into thinking he is a redhead when his friends dye his hair as punishment for stating that redheads are inhuman and have no souls. True to form, he switches his position and proposes that all non-gingers be exterminated. Had his plan come to fruition, the resulting gene pool would only contain homozygotes for MC1R ginger alleles. In other words, this would work.

Guardian, to support and assist in the repopulation of the ginger people of the earth, should this auburn apocalypse transpire, by maintaining the various alleles in subsequent generations through vigorous procreation. Naturally, I sought my wife's approval (after I had attempted to explain the extreme unlikeliness of this happening).

In fact, my services would not be further required at all, because I had already done it. When I had my genome sequenced, by BritainsDNA—the very same company behind this entirely specious extinction nonsense—there it was: MC1R Val60Leu. I carry a mutation for ginger hair. At position sixty in the MC1R protein, instead of the amino acid valine, I have a leucine. In the gene, a single letter of DNA is changed; sometime in my ancestry, a G became a C. This was partially surprising, as I've got black hair on my head (though it's flecked peppercorn these days; weary melanocytes give up the ghost as one ages),* no doubt from my Indian heritage. My father had dark brown hair (nowadays he's a silver fox), and his family, mostly from the northeast of England and Scotland, are to our knowledge largely ginger-free too. It's also not very surprising as Rutherfords hail from the northeast of England and Scotland, where redheads are unusually common. A few years ago, in the throes of capricious youth, I stopped shaving for a summer and sported a beard. In among the black was a crop of hairs that were the perkiest auburn, like poppies in a scorched field. I

* In March 2016, the first comprehensive study of the genetics of head and facial hair was published, and revealed a suite of alleles, of genes mostly already known to us, that associated with characteristics such as monobrow, density of beard, male pattern baldness, and tendency to go gray. At the time of writing, these variants were not yet added to the commercial genome testing databanks, though I suspect they will be sooner or later. The fact that the genes were known, and have multiple known functions already, further reveals that genes do many things in our bodies, and though my (or your) graying might be influenced by genetics, it is part of a typically complex interaction between biology and our environment.

plucked a couple and under microscope compared them to a friend's very obviously red hairs and they were indistinguishable. So it's there in my genome. My wife is blonde, and of my three children, two are blond, and two have brown eyes; my flaxen son has eyes so blue he could be Swedish. One of them may well carry a MC1R red allele. My wife's sister is a strawberry blonde, and has two children by a very ginger-haired man, and they have the most glorious bright red hair you could imagine. So, what are the chances of this oh-so-northern-European trait going extinct? Roughly, somewhere between none and zero.

Red hair appearing exclusively in beards is not uncommon, though we don't really know why. Forgive us; it's not really been a research priority over the last few decades. The ginger twin study showed that my mutation, Val60Leu, occurred most often in "fair/blonde and light brown" hair, three things I emphatically am not. That again is typical for genetics—presence or absence of gene variants are rarely fully absent or fully present in populations. Such is the nature of human variation: We're very variable. One version of the allele exists at a higher frequency in Irish redheads. One version has a slight association with a lower pain threshold. One version seems to subtly affect how an individual responds to an anesthetic that dentists frequently use. This is just one gene, with many versions, and it has a variable phenotype, though we can mostly see the outcome with fiery clarity. Even when we know the genome intimately, and the patterns of inheritance, and the history of the DNA, and the migration patterns of the people who carried it, and evolutionary pressures that led to the perpetuation of the genes and the phenotype—even when we know all that, how it manifests can still be mysterious and surprising. Anyone who says differently is selling something.

The British are coming

While we're in the northwest of Europe, let me indulge in some national pride to scrutinize this sceptred isle, and the finest genetic analyses of a people yet undertaken: Just who are the British?

Archaeologists sometimes use technological cultures as being definitional of people or eras. These are often broad and scattered characteristics incorporating a multitude of skills. The Beaker people, for example, made distinctive drinking cups and vessels over a couple of millennia beginning almost 5,000 years ago, and also had bronze and other metalworking skills, shot arrows, and were spread all over Europe, with many subdivisions across the land. Or before them there were the Linear Pottery people, who were similarly widespread and made pots (though not in a line: the linear name refers to bands of lines on these pots incised into the clay. It's not the most descriptive definition, but so it goes; that's what they're called). These are broad, sometimes woolly definitions that are helpful for us to see vast transitions in culture across large areas and huge populations. Within those potters were the farmers who edged out the hunters of the northwest in the Neolithic.

The tide of people in Europe has been incessant from the first humans onward. Although the sea is not a barrier to gene flow, continuous landmasses make for easier journeys, and the band of sea between us and the French, Dutch, and Scandinavians has slowed down a free flow across our borders over history. We've not been aggressively invaded since 1066. We've not always been an island nation, though. My homeland, East Anglia, was not the gentle coast of my youth: It was simply continuous land to what is now the Netherlands. This terra firma, called Doggerland, played host to the estuaries where both the Thames and the Rhine flowed out into what is now the English Channel. Today, its shadow is referenced in the *Shipping Forecast* on BBC Radio 4 as Dogger Bank, a 7,000-square-mile submarine sand mound that was once dry land. Twenty-one thousand years ago, during the period when northern glaciers covered these lands at their maximum girth, the sea was more than a hundred meters lower than today, and the land surrounding Dogger was solid too.

Across this icy bridge trudged mammoths and hairy rhinos, during and before the last Ice Age—and people. We don't really know who these folk were. Neanderthal teeth have been found as far west as north Wales. Other than that, the human remains are scant, though the detritus of their lives is easier to come across. There's Boxgrove Man, a 500,000-year-old forty-year-old *Homo heidelbergensis*, identified only by his shin bone, which has teeth marks on it, from either cannibals or animal scavengers. Around his last resting place on the South Downs are hundreds of Acheulian axe heads that span from 800,000 until his time. And there's Pakefield in Suffolk, which 700,000 years ago was in the middle of Doggerland, and was the home of an unknown people who also knapped blades out of the flint still common on the Suffolk coast today.

These old bones are almost always named something "Man"—Neanderthal Man, Cro-Magnon Man—which can be a bit tedious given that half of all humans have not been that, but the skull found in a pit near a golf course in Piltdown, Sussex, in 1912, was only half man. For a short while he was a celebrity, but turned out to be a shameless fraud, an orangutan's jaw jammed into an old but modern human skull, the work, probably, of Charles Dawson, who was so desperate to become a proper grown-up scientist, join the Royal Society, and gain international repute that he engineered this and other elaborate deceits over many years.*

* I wish he hadn't. Creationists frequently cite Piltdown Man as evidence of the imagined folly of human evolution, despite the fact that it was formally repudiated in 1953 and not a single scientist alive thinks it is real.

On the beautiful Norfolk coast, there's Happisburgh (pronounced Hays-bruh), a small village with a lighthouse, a church, a superb pub, and not much else, apart from almost a million years of human occupation. The North Sea grinds away at the sandy coast every year, and houses inch closer to toppling into the brown seawater below. As a result, the past just falls out of the cliffs, and pokes out of the beach. A Paleolithic axe head was picked up by a man walking his dog in 2000, and now the annual digs there recover more tools, and the remnants of hunting and butchery of bison, and the fragmented remains of rhinos, elephants, horses, elk, and a colossal beaver. And then in 2013, human ichnofossils were found. These are trace remains, the evidence of living activity in the absence of the bones themselves. In the clay bed only revealed at low tide, there are footprints. They look like they were made by a size 9 Doc Marten boot, but the dating of these beds says that those feet in ancient time walked upon England's mountains around 800,000 years ago. Toes and details are not present, as they've been weathered by oceans and time, so we don't know what species these feet were.* Nevertheless, Britain has been host to humans for almost a million years.

The ice retreated, and the seas rose. Some believe a mega-tsunami crashed into Doggerland and slid the land into the sea forever, and since 6500 BCE, just when the farmers were arriving in western Europe, we have been an island. That makes us interesting for both conquerors and geneticists. We put names to people with the cultures from archaeology that appeared to share cultures and geographical origins: Celts, Picts, Angles, Saxons. Some of them are better defined than others. The Romans at least came from Rome,

---

* Chris Stringer, the doyen of human evolution in Britain, thinks it likely that these feet belonged to *Homo antecessor*, a human that lived between 1.2 million and 800,000 years ago, known primarily from caves in the Atapuerca mountains in central Spain. He's usually right about these types of thing.

though for the 400 years that the Romans were in Britain most of them were conscripts from Gaul. For all the attempts to put discrete labels on the people of the past, genetics has a habit of defying them.

We haven't been significantly invaded since 1066, though immigration has trickled continuously since. Before that, though, there were plenty of invasions en masse, rulers who came and stayed, or came and went, and left their cultures behind. The question for the twenty-first century is whether they left their genes. This was the subject of a mammoth project called People of the British Isles in 2015, run by Peter Donnelly and Sir Walter Bodmer, two geneticists not primarily known for their interest in history, but in the genetics of disease, in particular cancer. But genomes are DNA, and DNA is data, so it's a short jump from the pernicious variation in a disease gene to the benign variation that tracks evolution and history. Of course, overall we are incredibly similar in terms of our genomes. But the levels of detail we can now extract from DNA, using delicate and highly perceptive statistical techniques, are unprecedented, and reveal secrets from the past undetectable by any other means—wispy fingerprints only noticeable with forensic analysis.

Two thousand and thirty-nine people were carefully selected across Great Britain, who spat in the tube and gave up their history. They had been chosen wisely: All had four grandparents who were born within fifty miles of their own birthplace, in order to exclude the waves of immigration from the colonies and elsewhere that came in the nineteenth and twentieth centuries. Many were of retirement age, so their grandparents were decidedly Victorian, which lengthens the odds of isolating DNA that has been as static as possible across the generations that span thousands of years. Six thousand Europeans from ten countries were the template from which they checked, looking for similarities and differences between the people of Britain and the people on the mainland.

The zoom you use on DNA determines the resolution of picture it paints. "Human" becomes "non-African" becomes "broadly northern European." Magnifying to the minutely fine-scale shadows of British DNA reveals a history of the people, not just kings and queens, but proper hoi polloi. This powerful lens is a great leveler, a historical source less tainted by victory or politics. The genome is a neutral record of sex, sometimes on a grand scale, and admittedly sometimes reflecting the politics of the time. The delicacy with which the traces of ancestry can be extracted can change, or reinforce history. These are patterns invisible but for the subtlest of statistics, long lost webs of family trees that are just threads of silk. From the spit in the tube, half a million individual spots on the genomes of the Brits were identified and fed into a computer, all individual SNPs, and neutral changes, meaning that they don't convey any significant advantage or otherwise to the owner. Each one was compared to each other one, and compared to each of the 6,000 European controls. This is history enabled by computer power.

The software shuffles people into groups as similar as it can, based on the 500,000 variables on each genome, but by design doesn't take into account geography. It was only asked to sort people who were genetically more similar to each other than to the rest of the sample, seventeen groups in all, all color coded with a symbol. Then, when the sorting was done, Walter Bodmer's team placed each grouping on a map of Britain according to the location of their grandparents.

Only at this resolution are they revealed, but the differences are stark. If we were all completely homogenized, then the map would look like pebbles on a beach, and with fewer people in the samples, that is what the map would've come out like. But it didn't. The ancient genetic map of Britain shows clear clusters of people who were more wispily similar to each other than to the rest of the

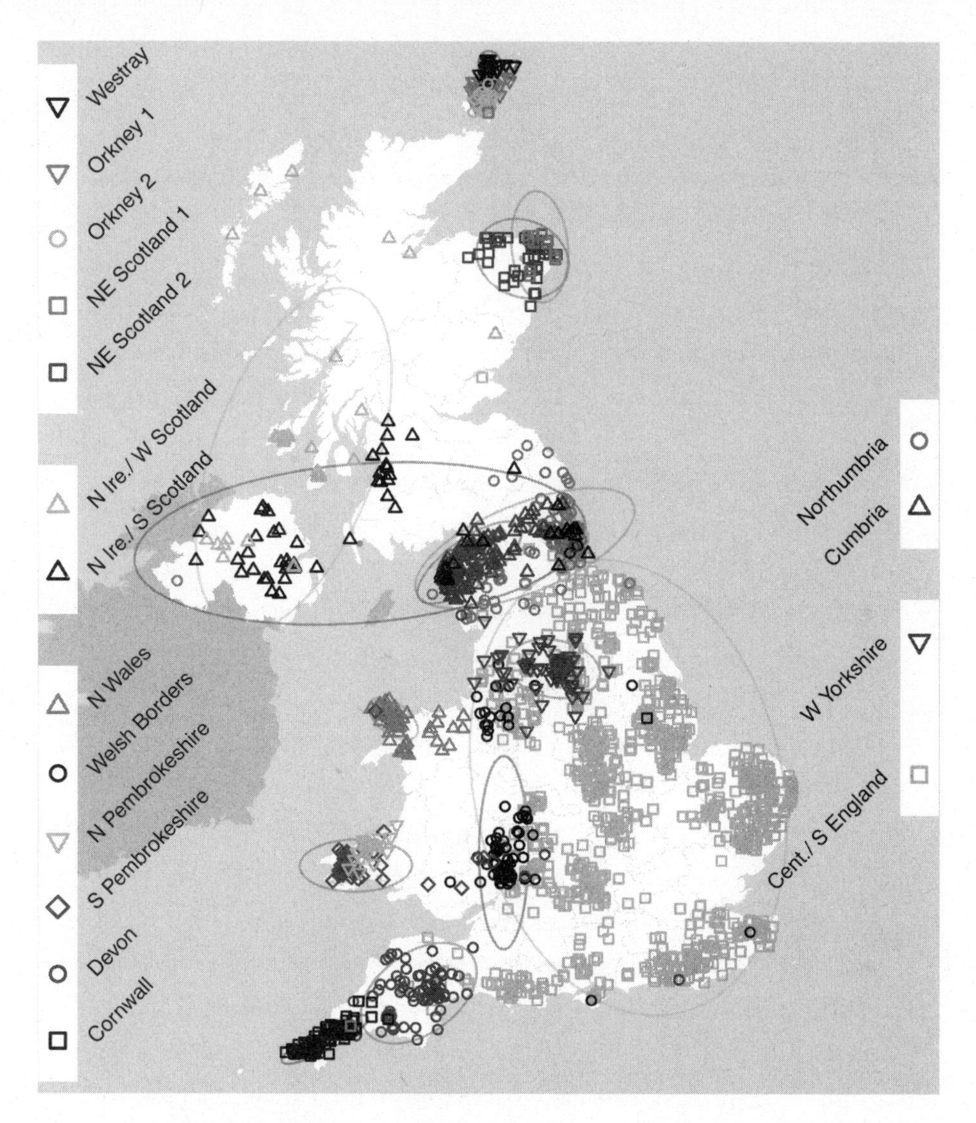

The fine-scale genetic map of the people of Great Britain.
The People of the British Isles project sorted the DNA from 2,039 Britons into genetically similar clusters, and placed them on a map according to where they were born. The map reveals local constellations that reflect the history of the British Isles. (© Stephen Leslie, University of Melbourne, Australia. Based on data published in *Nature* 2015. Contains OS data © Crown copyright and database right 2012. © EuroGeographics for some administrative boundaries.)

country. The differences are slight but striking, and only visible en masse.

The most genetically distinct are the people of Orkney—a cluster at this level of detail that cuts them off from the rest of us. Maybe that is not so surprising as Orcadians are an island people on their own, and remote even by Scottish standards. Around a quarter of their genome appears to come from Norwegian Vikings, as precisely as 1,100 years ago (assuming twenty-nine generations based on the number of genetic changes, and a generation being twenty-eight years). That places an invasion from Vikings in 916, who stuck around and got on handsomely with the locals.

At this resolution, the Cornish are distinct from the people of Devon, and the borders of this genetic difference are very close to the county lines, reaching from Bodmin Moor to the Tamar Estuary. The Welsh, they look distinct from the rest of Britain too, and within Wales, the north and south settle apart as well. Even within southern Wales, the divisions of culture are borne out in the genome. An invisible line slices through the southernmost tip of the Gower Peninsular, at the bottom of Wales. It bisects Pembrokeshire to the point where that county meets its neighbor to the east, at the River Taf north in Carmarthenshire. Since the sixteenth century this domain has been known as *Anglia Transwalliana*—Little England Beyond Wales—and the boundary called the Landsker Line. It's a language barrier more than anything else, and has no formal or legal status, though it has lasted a thousand years. To the south is a pocket of English speakers, whereas the rest of Wales spoke Welsh (at least historically; English is spoken everywhere today). It's drifted over the centuries, a few miles north or south depending on the local movement of people and traditions, but largely it's preserved as a vague border a few miles deep. There's a neat-ish line of forts that map roughly along the line too, Norman castles from Roch on the west to Laugharne on the east, but it's a stretch to say these were frontiers protecting this informal border.

The People of the British Isles study showed that the Landsker Line is there in the genome too. Residents of *Anglia Transwalliana* leapfrog over 150 miles of south Wales to share more genetically with the English. Earlier genetic maps of the Welsh had not spotted this and had shown no significant difference between the Welsh as a whole and the English, which is no fault of those genetic cartographers; it's simply that the latest techniques can pick out these differential ghosts from a genomic mass of similarity.

These are the finest of differences, but they show the remnants of peoples who have, on average, stayed put rather than nomadically roamed. It won't surprise the people of West Yorkshire that they pop out on this map as being distinct from the surrounding people, a tribal gathering called Elmet. Nor people of Cumbria, which overlies a Celtic settlement called Rheged. The map says profound things about history too. As well as the shadows of regional populations in the UK, the comparisons with European DNA show where bits of the modern British genome originated, or didn't. There's very little ancient Italian. The acres and tons of archaeological evidence for 400 years of the Roman Empire in Britain are rich and wondrous. But the biological traces are comparatively absent.

The Celts are a people whom we romantically think of as being tough Welsh, Scottish, Breton, or Irish, with a style of art and culture instantly recognizable with often abstract twisted shield, bolts, and crosses. They slide up the west coast of Britain, from Cornwall, through Wales, and into Scotland, Picts to the north, Saxons to east.

But they're not a cohesive group of people at all. According to the British genome, Scottish Celts are more different from Welsh Celts than either are from the English. The same goes for the Cornish, who resemble the Breton Celts 250 miles to the south in France. Fiefdoms that we refer to as Celtic stick out, such as Elmet in the West Riding of Yorkshire, Rheged in Cumbria, and Dumnonia in Devon and Cornwall, but are dissimilar to each other at this genetic resolution. Similarities seen in cultures, such as farming techniques, or the patterns on decorative trinkets may be common, but the DNA shows that they are not necessarily rooted in shared ancestry. They may have traded at some point, or copied, or it might be coincidence. But, according to the genetics, there wasn't a point where a group of genetically similar people spread into the extremities of the British Isles and settled into a culture that we now call Celtic. That word is a modern invention of a presumed people that isn't reflected in Britain's DNA.

Wave after wave of invasion until the Normans came shaped the cultures and histories over centuries. Comparing the genetics of the British now with Europeans today can tell us who came and, via close relations with the indigenous residents, left the biological signatures of their visit.

The south is the largest solid block of color on the genetic map, though; a bent rectangle roughly cornered by Newcastle, East Anglia, Kent, and Dorset. We call them the Anglo-Saxons, but they didn't. They were more like a scattering of peoples from northeast mainland Europe comprising tribes such as the Mierce, Gewisse, or West Seaxe. They followed, or maybe contributed to, the decline of Roman Britain, coming from the countries that are now southern Norway and Denmark and Belgium in the fifth century, maybe migrating away from the influx of people from the east, such as Huns and Bulgars and the unexcitingly named Alans. Following on from the Romans' *exeunt*, many theories have been proposed over the years of what happened when we plunged into what is daftly called the Dark Ages—as if Rome was the light, and a shroud fell over Britain as sophisticated culture retreated from our shores. It is true we didn't write down as much after they skedaddled, but culture abounded, as did agriculture, and historians continually debate how good or bad things were. Some theories suggest that warmongering Saxons turned up and wiped out the abandoned natives, or freed them from imperial tyranny. The genomic map showed a solid block of red that corresponded to the areas that Rome ruled and that the Saxons inherited. It shows admixture, a virtually even spread of genes from older Britons and from European Saxons. They came, they saw, and they just integrated into the lives of the existing population—migration, intermarriage, and assimilation. Unlike the Romans, they stayed, and their genes have endured ever since.

One thing is certain in human history: Nothing is forever.

In the ninth century, the Vikings came from the north. The Danish started bothering our shores in the 800s, with frequent pirate raiding parties to the east coast. They came again with intent in 865: An army led by Ivar the Boneless, Sigurd Snake-in-the-Eye, and Halfdan Ragnarsson arrived in Suffolk to conquer the Anglo-Saxon territories. They marched north and took York and Cumbria, and over the next few years marched back down to London and conquered East Anglia too. Danelaw was established, a ruling body and land that blocked out most of England, with Wessex under Alfred to the west. This territory included England and Scandinavia, with undulating borders and politics, and an influx of Norwegians, until 1066, when Harald Hardrada got an arrow in his throat at the Battle of Stamford Bridge. Harold Godwinson took the English reins for a short spell, until he got an arrow in his eye from a Norman conqueror down south in Hastings later that year. And thus ended the permanent residence of the Vikings in Britain.

Yet there is virtually no trace of the Danes in the British genome. Compared to the Angles' and Saxons' and even the Norwegians' genetic legacy in the north of Scotland, there's an absence of Danish DNA despite a long adventure here. This says something about their 200-year rule. They didn't integrate. They may have shaped the lands and defined and defended borders, and given us the days of the week (and hundreds of other words: "a berserk freckly husband is a blundering guest in hell"), but they don't appear to have left any distinctive DNA. The Vikings might not have been as bloodthirsty as popular depictions; they farmed and wrote and created great art. Their first interactions with the Brits were piratical parties, with assumed rape and pillage, yet the former left no genetic trace. Like the Romans before them, it seems that those Vikings wielded their power from above, absolute Cnuts and Haralds ruling from the top down, with no enduring relations with their subjects.

In earlier and other chapters, I've gone to lengths to explain how we are all descended from a very small group of people, and that all lines of ancestry cross in the surprisingly recent past. You are descended from Vikings, because everyone is. If it now seems like I'm separating all these pockets of Brits out as distinct, it's a facet of the zoom, and the People of the British Isles project doesn't negate that. Because every person has so many lines of ancestry above them, a branch from everyone will cross at some point. But you still may well have more branches from Vikings than from Angles, and indeed, more lines from Norwegian Vikings than from Danish, who didn't seem to leave many British descendants at all, as they ruled from afar, and like the Romans before them, didn't get stuck into the locals much. And although this sifting puts the Welsh as distinct from the rest of the British Isles, it also mocks any modern nationalistic pride that you might derive from that. The north Welsh are as different genetically from their southern compatriots as southern Englanders are from the Scottish, or the people of

Devon from the Cornish. Even if you're more Welsh than your neighbors, it's only in the details, and you're still part Viking, Saracen, Angle, Saxon, and, as we shall soon see, part Holy Roman Emperor.

This map of the British Isles is unprecedented. It ties history and people—*the* people—the normal folk who arrived and traded, settled and integrated in farms or homesteads with people already there. It shows a picture of simultaneous homogeneity and diversity across our lands. What a beautiful map it is too; to see those symbols cluster according to ancient lands is invigorating and reassuring. We're all the same people, we're all slightly different, and the differences show up in funny pockets, reflecting a land that has such a profound and rugged history, but that has been pretty stable now for centuries. I think of my children, mutts with their genomes from south Wales and Ireland from their mother, and from the northeast and Scotland from me, plus a dollop of south Asian to spice things up. They are our twenty-first-century national dish: a very British curry. The same techniques can now be applied to us in the postcolonial era with the rich flow of the peoples of the world that now make up our countries, and it can be rolled out across every nation on Earth.

930 CE, Bláskógabyggð, Iceland

There are other strange tales further to the north. Sixty years after people first set foot on Iceland, the baron of Thingvellir murdered a local. Soon after, he stood trial before a court of his peers above a wide volcanic plain. As punishment, Thingvellir's land was seized by the local families, and they decided to exile him, not abroad, but to the interior of their island. There, in its barren, lunar terrain he would surely live out his days quickly. The new common land became the site of the world's first national democracy, grounds set for assembly and lawmaking. The Icelandic parliament has now moved into the capital, Reykjavík, but it is still called the Althing. Thingvellir's bleak, brutal but green land (when not blanketed in snow), perches atop a jagged cliff overlooking the mile-wide valley that separates the North American tectonic plate from the Eurasian one. Iceland straddles a rip in the earth's mantle, and the two plates riven through the land are slowly pulling it apart, at the speed of fingernail growth. There is the source of Iceland's living rock, still angry to this day—Eyjafjallajökull and Bárðarbunga, and a smattering of other minor volcanoes blowing their tops in the last five years.

Iceland is a weird place. The living landscape is virtually extraterrestrial—glacial and barren, volcanic and lunar, and its culture and language rich and bewitching. Its latitude means that the winters are black all day, and the summer sun barely dips below the horizon. Even the neighboring Scandinavian countries think the Icelanders are a bit odd. The country has an extremely uncommon history, which makes it paradise for a geneticist. It is my second favorite country on Earth, other than the land of my birth.* The tales of the first families are logged in the Icelandic Sagas, collectively one of the most important European texts. Written over a sixty-year period, they contain events of world-changing significance, such as the first steps of European feet on North American lands, and plenty of ghosts and trolls, sex, violence, drinking, and subsequent puking, as surely befits the Viking reputation.

---

* I am not the first Dr. Adam Rutherford to comment on the founding of the Icelandic nation. My namesake was a well-respected twentieth-century scholar of Egyptology. Twice I've been mistaken for him, which is both physically and intellectually disappointing, given that he a) has been dead for decades, and b) was a devout Christian biblical literalist. In 1937, he wrote a pamphlet detailing the foundation of the Icelandic people not being the tale of the Sagas, nor of modern genomics, which were of course then unknown. Instead, as part of his misguided attempts to evangelize his fundamentalist Christianity, he attributed the Icelandic origins to the twelfth tribe of Israel, that of Benjamin. Alas for that Adam Rutherford, evidence for Icelandic trolls is more robust.

The genealogy of virtually every Icelandic settler is recorded. Garðar Svavarsson was probably the first,* lost in the north Atlantic in the early 860s, but he only stayed a winter. Flóki Vilgerðarson was next according to the *Landnámabók—The Book of Settlements—* which documents Iceland from first residents to its time of writing in the twelfth century. Ingólfur Arnarson and his family arrived in 874 CE, and took their place as the first permanent residents in the southwest fjords, a place he called Reykjavík—the Bay of Smokes. Other settlers followed that year, and the *Landnámabók* lists more than 400 families. By the formation of the Althing in 930, according to the *Landnámabók*, there were 1,500 farms and place names and more than 3,500 people.

Island populations often breed strange and interesting genomes. Because Iceland is so small, with a precisely known chronology of inhabitation to the present day, we have the most comprehensive log of everyone who has ever lived there since the ninth century. There have only been thirty-five generations of Icelanders. Since the end of the era of settlement, there's been very little immigration into Iceland of note. The population has never topped 400,000. These are all matters that make genealogy and genetics significantly easier. The Icelanders knew this, and decided with great foresight to add DNA to their database of identity. Theirs is now the most comprehensively studied genome of any people, both ancient and modern.

Since the beginning of the twenty-first century, the genetic genealogy of Iceland has been explored. As often happens in this business, the mitochondrial DNA and the Y chromosome were

---

* There were almost certainly people on Iceland before the Norsemen arrived, probably Gaelic monks, who didn't stay, and maybe left no descendants, as monks are supposed not to. A cabin might have been occupied in the eighth and ninth centuries, but was abandoned by the time the Icelanders arrived to stay in the 870s. It is their story that is enlightened further by the recent forays into the genome, and so I must set aside those earlier monkish wanderers.

first, as they leave the most obvious genealogical tracks. And those tracks are again unusual. Two thirds of the mitochondrial types are most closely associated with Scotland and Ireland. The rest and the vast majority of the Y came from Scandinavia. With little immigration, and a perpetually small population, these numbers give a much better handle on the origin of these DNA types than in admixed countries like the UK, or continuous landmasses such as Europe.

Consider what that means. Male Scandinavians bred with Scottish and Irish women. In the People of the British Isles project, the Orcadians came out as the most different.

Norwegian Vikings came to those beautiful islands repeatedly, but unlike the Danish in Britain, enjoyed local hospitality and left their mark very clearly. Scotland and Ireland were known destinations of the Norway and Denmark Vikings' seasonal raiding expeditions. They clearly ventured further west and north too —the men, that is—and acquired wives en route. Whether those acquisitions were forced or voluntary is unknown.

That's the story based on Iceland's contemporary DNA. But with such a short history, the past was not far behind. Over the last couple of centuries, Icelandic archaeologists have dug around in the volcanic soil looking for the burials of the first wave of immigrants. In 2009, the teeth of ninety-five Icelandic settlers from the tenth century were wrenched from their jaws, and gave up their DNA to the small army of Icelandic geneticists led by Kári Stefánsson. He had traced his own ancestry back to 910 CE, to a warrior poet described in the Sagas as famously unattractive.

The graves are scattered around the island, many in and around Reykjavík, some to the north, but mostly on the periphery, as the interior is an unforgiving dominion. These were people who are thought to have died before 1000 CE, including probable jurors at

Baron Thingvellir's murder trial. The National Museum's skeleton collection hosts 780 former Icelanders, but most of them haven't been carbon dated, so their ages are not well documented. There's a neat historical cheat code here though, which is that Iceland officially converted to Christianity around 1,000 years ago, and burial practices changed as a result. The Norse buried their dead in north-south orientated graves around the homestead, mostly as singles, sometimes in small family plots, and included weapons, jewelry, animals, or boats that might serve them on their way to afterlives in Helgafjell, Hel, or Valhalla. Christians like cemeteries and east-west graves with none of the fun stuff.

The DNA extracted from these settlers confirmed to a degree the earlier studies of contemporary mtDNA, but was different in a very interesting way. The regions of the ancient mitochondrial genomes they looked at were more similar to contemporary Scots, Irish, and Scandinavian mtDNA than they were to contemporary Icelandic. This is a facet of the phenomenon of genetic drift. This occurs when the sampling of a larger population is not representative of the overall population. Imagine you have a bag of a hundred marbles, fifty blue and fifty green. If you sampled eighty, you'd probably get close to forty of each, and the ratio would be maintained if you were to seed a new bag based on those eighty. If you only took one, though, you'd have a 50:50 chance of getting a blue or a green, and if that was the basis of a new bag of a hundred marbles, suddenly, you only have blues. This effect is seen frequently in evolution when a founding population is small compared to the people from whom they came. This is called a founder effect, when their misrepresentative genomes become the basis of an entire population. In small groups it happens faster, and that is what we've seen in the Icelanders. Their founding genomes have remained similar in the countries whence they came—the north of the British Isles and Scandinavia. But there on the island, with no significant input from other sources for more than thirty generations, the genomes of the Icelanders have quickly drifted away from their founding mothers.

The Icelandic interest in genetics has moved on apace as the genomic landscape has evolved. Kári Stefánsson and his company deCODE are the driving force behind all things Icelandic and genetic, and they've set the bar high for the rest of the world. In 2015, they published the most complete set of the genomes of a people yet, the complete genetic readout of 2,636 Icelanders,* and sampled more than 100,000 others. This scale reveals patterns unseen in individuals, and so many rare or unspotted disease-causing or disease-related genes were found. Patients were found with a version of the gene MYL4, which causes an irregular heartbeat called early onset atrial fibrillation. They found versions of genes that trigger gallstones, others that strongly associate with Alzheimer's disease. A very odd thyroid condition showed its hand in a mutation that boosted a hormone if it came from the mother, but lowered it if from the father. That is difficult to explain, but it highlights the complexities of human genetics that we are only just beginning to understand.

They saw something else too: a very high level of genes that just don't work at all. Many mutations in genes just subtly—or in some cases profoundly—change the shape of the protein they encode, and therefore the way that protein behaves. Those small changes account for much of what we see as human variation—red hair, blue eyes. Some mutations introduce a termination message before the end of the protein, kind of like a period before the end of the sentence. Some insert extra letters in such a way that the whole protein

* They are now, I am told, up to 10,000 complete genomes, though these have not been published yet.

is out of frame; I compare this to a 35mm film running out of sync with the projector gate, so you see half the picture you're meant to—they're called frameshift mutations. If I only use three- letter words, it demonstrates the problem:

the big red fox ate cat pie

Inserting a random letter (x) near the beginning, but maintaining the three-letter structure really messes things up:

txh ebi gre dfo xat eca tpi e

Versions of each of these can also simply make a protein functionless, out of step with its unmutated code, and these are called "loss of function mutations" or LOFs. deCODE's big 2015 papers showed that the Icelanders have way more of these than most people. One in twelve of their population has inherited a LOF. Across the population sampled, they found 1,171 of these totally nonfunctional genes. These were found in living people, so they are clearly not mutations that are not conducive to life, nor fatal during development. We have 20,000 genes or thereabouts, so you might think that missing one or two might not be such a bad thing. But it really depends on which one you are short of. The 1,000-plus identified in Iceland are the starting point for working out not just what each of these genes does, but what redundancy is built into our biology such that losing whole genes does not prompt loss of life.

Life on a small island

And perhaps unsurprisingly given the island, and the small founding population, they found a high degree of homozygosity—inheriting the same gene from both parents, and thus increasing the likelihood of a recessive condition appearing. This is a facet of inbreeding, of what geneticists call a "small effective population size." With such a persistently restricted population, incest is a constant specter, and we know how that can affect families (see Chapter 4, part 3). They know it too, and caution abounds. Whereas in a small town with typical patrilineal naming, you might raise an eyebrow if being chatted up by someone with the same surname, the Icelandic naming convention does not volunteer a meaningful alert to possible genetic proximity. The father's first name becomes the root of the child's surname, with daughters adding -dóttir, and sons adding -son. So for example, Iceland's most famous child is surely the singer Björk,* whose surname is Guðmundsdóttir, the daughter of Guðmundur Gunnarsson, son of Gunnar Guðmundssón. In 2013, software writers Sad Engineers Studios created Íslendinga-App. It's effectively a database of most of the people who have ever been recorded as having lived in Iceland, drawing its data from the Íslendingabók genealogical library built from records dating back more than 1,000 years. Two users can link their apps on their smartphones together to see how they are related, and genealogical proximity sets off the Sifjaspellsspillir alarm—the incest spoiler.

What a strange land, and a wonderful people. If there is a nation on Earth better studied, from their origins right down to their DNA, I am unaware of it. Compared to the rest of the world, their roots are not deep, and as a result their genetic diversity isn't either. But their records certainly are. Sometimes a great lack of diversity can be fruitful, but only if you've got the foresight and imagination to make it flourish.

* Who is an incomparable genius.

We all fall down

Let's leave the human genome for a moment to dig into one that has shaped empires, culture, and devastation across Europe and beyond. It's not just our own DNA that tells us our history. *Yersinia pestis* is a fairly typical rod-shaped bacteria, a single cell around two micrometers in length (ten times less than the width of a typical European head hair), and stationary with no propeller or other means of movement, as some motile microorganisms have. Yersinia is unremarkable—it was discovered by Swiss-French bacteriologist Alexandre Yersin in 1894, and acquired his name in the 1960s.

Now, imagine you are *Xenopsylla cheopsis*, an Oriental rat flea, tucking into a typical meal by biting an animal, say, a species of fat Russian squirrel called a marmot. You have two sawlike laciniae, which lacerate the skin, and in between them is the hollow needle we call the epipharynx. Together these form a channel through which your saliva can run into the victim, and blood is sucked out. Itchy and annoying to the marmot, but that's how a flea gets its fill.

All animals are replete with bacteria, inside and out. Most of these organisms do nothing harmful, and many are positively useful. Animals are all different though, and a marmot is different from a flea is different from a human. Yersinia can fester in marmots with no particular effect, and does so to this day in the prairies of the Asian steppes. As a voracious flea, you suck up some Yersinia cells, which pulse down into your digestive tract along with all the rest of your bloody meal. In the warmth of your mid-gut, the bacteria start to multiply, and produce proteins that help them coalesce into lumps with clotted blood, and acquire a thin membranous skin to hold them in place. This makes it difficult for you to digest the blood you've already consumed, so you, the flea, still feel hungry and hop onto another animal and try to get some more food. This time, the fresh blood you suck floods into your mid-gut and dislodges some of the lumps, filled with now rampant Yersinia, which rush back down your laciniae, and into the poor cretin you're chewing on. Yersinia might not have legs, or even the whiplash rotor that many bacteria wield. But it sure knows how to get around.

This is not a particularly unusual life cycle of a parasite. Every known organism has parasites, and the relationship between host and predator has been a major driver of many aspects of evolution. But if that second critter that the flea is trying to get some nosh out of happens to be not a marmot, but a large bipedal mammal of the species *Homo sapiens*, then what follows is seriously bad news. Your skin is normally the first defense against infection, but Yersinia has already breached that by persuading the flea to try to pierce the skin to feed, and then regurgitating the pathogenic cells straight inside. Once there, it starts doing a whole range of things to further its own life cycle, and in doing so, end yours. It switches on genes that make proteins that stick to your epithelial cells, that is, cells that are on any wet surface—in your gut, in your mouth, lining your blood vessels—and invade them. It makes proteins that render it impervious to phagocytosis—the process where your large hungry Pac-man immune cells called macrophages literally swallow invaders and digest them. It produces proteins that punch small holes in your cell membranes, and these allow the bacteria to seep out to infect more cells. Yersinia blossoms and proliferates in lymph nodes, where it can evade our highly evolved immune system defenses with an arsenal of nefarious tactics. The host cells rupture, and spewing their diseased guts into the intracellular milieu invokes the classic signs of an inflammatory response: *tumor*, *rubor*, *calor*, and *dolor*—swelling, redness, heat, and pain. Some of your metabolic pathways are ectopically activated by Yersinia, which prompts the macrophages to submit to the persuasive will of their invaders and commit suicide.

Cellular self-destruction is a healthy normal part of living, as cells have completed their use and need to be disposed of, such as in the webbing between an unborn child's fingers. But this suicide is not part of the body's plan, and absolutely unwelcome. The macrophages are wiped out, and this has the domino effect of weakening your immune system further. Pain and fever become crippling headaches and your body is plagued with swelling. Cell death in blood vessels denies the oxygen and nutrient supply to the extremities—and as the cells of your toes and fingers die, they become necrotic, blackened, and seeping with pus as gangrene sets in. Your lymph nodes swell like purple water balloons—buboes—notably in your armpits and your crotch, but by this stage you're too feeble to do anything much. You are riddled with the bubonic plague, and a fortnight after that hungry flea puked *Yersinia pestis* into your bloodstream, you're dead.

You might get it in your lungs, and cough and wheeze and spread the airborne bacteria to other people.* Or it might infect your bloodstream, with a host of other cellular apocalyptic traits, and as with bubonic and pneumonic, septicemic plague will end you, if not treated. Nowadays, we have good treatments, antibiotics that will block Yersinia's ability to reproduce, and if taken in the first few days, plague is easily manageable. Nevertheless, outbreaks do still occur—as in Colorado in the United States in autumn 2015. These days they are restricted to individuals, and quickly identified, and subsequent quarantine and treatment with antibiotics stops the spread dead.

---

* It's often said that the nursery rhyme "Ring a Ring of Roses" is a plague incantation, with the roses being reddened marks preceding buboes, "a pocket full of posies" some kind of traditional herbal protection and, of course, "we all fall down" (and also possibly "ashes, ashes") being representative of an inevitable death. However, this is almost certainly a twentieth-century post hoc analysis, and is rejected by most academic folklorists.

Bringing up the bodies

In the gloaming of the Roman Empire, matters were much worse. Rarely has a disease altered the course of a continent's history so profoundly, but *Yersinia pestis* managed it at least twice. The first decimation occurred in the sixth century, the epicenter of the infection being Constantinople. There, Justinian ruled the Byzantine Empire, including during a plague that lasted just a year, in 541, but at its peak was an apocalypse. The historian Procopius details some of the devastation of Constantinople in his book *Secret History*, written at the time. He suggests that when it was at its most virulent, 10,000 people per day were dying, though modern historians estimate a more conservative but no less baffling 5,000 every day. Over this bout of infection, something like 25 million died across the empire. Numbers are difficult to verify at that time, but with the waves that followed in the next couple of centuries, this number is estimated to have doubled. This pandemic extended its necrotic fingers all over Europe, to Germany, France, Italy, Spain, and even North Africa. Historians speculate that the obliteration of so many souls over the waves that followed for the next two centuries made significant contributions to the end of Rome's rule, and plague-weakened armies opened the door to people of the Middle East.

Procopius describes plague victims in a port near Suez in Egypt in 542 with the classic symptoms seen during the Black Death a few centuries later (which we will come to in good time):

> A bubonic swelling developed; and this took place not only in the particular part of the body which is called *boubon*, that is, "below the abdomen," but also inside the armpit, and in some cases also beside the ears, and at different points on the thighs.

> . . . Upon opening some of the swellings, they found a strange sort of carbuncle that had grown inside them. Death came in some cases immediately, in others after many days; and with some the body broke out with black pustules about as large as a lentil and these did not survive even one day, but all succumbed immediately.

This is the first description of what has been presumed to be that most virulent and fatal form, bubonic plague. Where there is corn, there are rats, and he suggests that trading corn with Egypt may have been the source of those first infections.

Genetics says otherwise. Work has been feverish in the last few years, as ancient genetics has found its feet. The dead are abundant when it comes to the various plagues, and we've acquired plenty of samples of *Yersinia pestis* from all over the world now, and taken out their genetic code. Nineteen teeth from twelve sixth-century plague pits in Aschheim near Munich provided the source of the Code of Justinian. In among the ancient human DNA are the remnants of other species that loiter around our bodies. A 2013 study ground out DNA from those teeth and found without doubt the same *Yersinia pestis* we see today. This had settled a long running debate about whether that great plague was in fact bubonic. But it also allowed us to place the genes of the Bavarian graves in an evolutionary tree, just like we do with humans and any other species. The similarities were unequivocally eastern, and not African as several historical sources had suggested. It appears that the trade from China, possibly along the Silk Road, had ferried these terrible bacteria from their benign existence in the east. The reservoirs of *Yersinia pestis* today are in rodents in the steppes, and, though this is speculation, it is easy to imagine that these might have been the source of the fleas that were harbingers of death for so many millions.

The Black Death

And it came from the east a second time. Halfway between two London icons—Smithfield's meat market and the theatres of the Barbican—is the East Smithfield Black Death cemetery. This part of town is now lush, salubrious, and fashionable, but under the surface there is little but death. A two-hectare site in one of London's typical squares has been excavated in the years since 1986, subducting its edges under the Royal Mint. Scraping away the veneer of august London prosperity revealed a horror story. *Inhumation* is the word we use to describe the burial of the dead, and between 1348 and 1350 there was mass inhumation in London. The Black Death had made its way from mainland Europe to Britain, where by 1350 it had cut through a third of the population, and this site was annexed as the first plague pit.

It's a human dumping ground, a midden more than a cemetery. Christian burials were a social necessity in those days, but people were dying at a rate where the living were barely capable of burying the dead. Some estimates suggest 200 people per day, and so East Smithfield was hastily designated a place for inhumation of thousands and last rites on an industrial scale. It's known as a "catastrophe cemetery," and of the 600 bodies catalogued so far, some are neatly lined up, others strewn willy-nilly, tossed to their grave in haste. The demographics of who is in the pit are way out of kilter with normal patterns of death in the fourteenth century. Mortality typically picks out the very young and the very old. In the plague pit, around a quarter are children, and most of these younger than five years old. But the rest are adults, and most of these are younger than thirty-five. These numbers show that the plague struck hard and fast, and wiped out a large proportion of an otherwise healthy population.

Johannes Krause pulled DNA out of this pit, but he was after the bug, not the people. By probing the full sequence of the London plague for the very first time, Krause and his team unearthed the evolution of *Yersinia pestis*, and the genomic tracks of its terrible journey. An earlier study had shown that, just like the Plague of Justinian, the Black Death in the 1340s had also originated in China. With a publicly available database of the full sequence, the history and the genetics can be aligned. Over a five-year period we can track a course from Russia to Constantinople, to Messina, to Genoa, Marseille, Bordeaux, and finally London. All these ports acted as points from which radiation of the plague could crawl inland. En route, it claimed the lives of some 5 million people.

Just as in the Byzantine Empire 600 years earlier, wave after wave of outbreak crashed into Britain's population in the centuries after the fourteenth, and it was only after the Great Fire of London in 1666 that this pandemic was crushed. Krause's work also shows that it never really went away. The pandemic might have ended, but the strains of Yersinia that cause bubonic plague outbreaks to this day are identical.

Let's return to human DNA for a moment. Across Europe, the Black Death left its dark mark, in the buildings, in the churches, and in the wretched huddled masses in pits below our feet. When a force of nature is this powerful and this aggressive, it can leave its signature in our DNA too. In 2014, an analysis of some of the genes in our immune system showed the hallmarks of evolutionary selective pressure in Europeans, including the Roma, Europe's largest ethnic minority. (Britons and Americans have traditionally called the Roma "Gypsies," but this is often considered derogatory.)

The origin of the Roma is northwest India, though populations have migrated into Europe on several occasions, and settled permanently in near eastern Europe in the eleventh century. This is an unusual and useful population structure for the archaic geneticist, as there hasn't been much admixture between the Romanian Roma and other Romanian Europeans. And for at least a thousand years, both populations will have endured the same environment, the same climate, and the same evolutionary pressures.

Of these forces of nature, the Black Death was a big one. An Indo-European union of scientists used the Roma to scan for the imprint of plague on our DNA. Mihai Netea from the Netherlands, working with Romanian, Spanish, and Indian scientists, sifted through publicly available genome databases and looked at the myriad immunity genes that we possess, looking at 200,000 SNPs—those individual variations in spelling that make up much of the difference between you and me and everyone else. In this pool of Romanian Europeans and Romanian Roma, they saw evidence for a cluster of genes undergoing evolution by positive selection. It's a process called a "selective sweep," and is described in more detail in Chapter 1. This signature was absent in people of northwest India whence the Roma once came (and also absent in Africans and Chinese). The presence of positive selection and lack of mixing of genes between the Roma and the Europeans is a telltale sign that these two peoples underwent a similar evolutionary force in the last 1,000 years, and that same force was not experienced in India. The Black Death never struck the subcontinent with the same vigor with which it blew through Europe, and the burden of mortality can be a great driver of evolution. Whether that is a result of chance or some innate immunity of those people is unknown.

The results indicate a strong association, but Netea's team back it up with some neat fine-tooth combing. Within the region of DNA that was apparently undergoing evolution in response to the plague is a family of genes with typically clunky names.* They're

* Drosophila researchers have no such formal nomenclature compunction.

called Toll-like receptors, or TLRs, and the proteins they encode sit on the surface of immune cells such as those hungry macrophages and sentinel cells. There they vigilantly await the advent of microbes with very specific markings. Upon identification of such an invader, an immune klaxon goes off, and the innate army of cells that protect us from within is activated. TLRs 1, 2, 6, and 10 are the combination that recognizes *Yersinia pestis*, and there they are, as a cluster, sitting right in that zone that has subtly but measurably evolved in Romas and Romanians, but not in the rest of the world. The mark of the plague is in our genes.

Procopius' *Secret History* describes the Justinian bubonic plague. But as we get better at reeling in the ancient DNA from out of old bones, the forensic work in piecing together the past gets deeper and richer, and tells us a different story. The latest analysis in this cycle of death was published in autumn 2015 by the Danish geneticist Simon Rasmussen, after his team had gone on a plague fishing expedition. Unlike the previous efforts that had looked in known plague graves for the genetic shadows of the bacteria, they simply surveyed the genomes of 101 people from graves scattered around Europe by pulling out the soft core of teeth from Bronze Age tombs. This technique demonstrates quite elegantly how the game of ancient genetics has changed in the last few years, as they effectively were trawling in a pool of 89 billion fragments of raw DNA.

Many genes are initially discovered in these flies because we can manipulate and mutate them so much more easily than in bigger animals, and with fewer ethical barriers than necessarily exist for humans. Often, though, mammalian geneticists would then look for a similar or equivalent gene in mice or humans and, more often than not, the name for the human equivalent would be derived from the original—and often descriptive or just plain fun—moniker given to the fly. The stellar Nobel prize–winning embryologist Christiane Nüsslein-Volhard discovered the Toll family of genes in drosophila in the 1980s, and the name was given after she was heard to exclaim, "*Das ist ja toll!*" which in English means, "That's fantastic!"

On a fishing trip like this, scientists need to know what they're looking for, and out of this infernal mess of DNA fragments, much of it contamination from other organisms, Rasmussen pulled out the last remnants of the genome of *Yersinia pestis* in Poland, Russia, Estonia, and Armenia.

Rasmussen's team decoded these bits of Yersinia genome and compared them to other strains, and to another closely related species, *Yersinia pseudotuberculosis*, which is not so malevolent and lives in soil. These comparisons set a molecular clock for when all of these versions diverged from each other,* and calculates the time when a most recent common ancestor (MRCA) existed. The number comes out at 5,783 years ago, thus predating the earliest reported infections by many millennia. That doesn't necessarily mean that we were infected in those ancient times. But among all that genetic debris, Rasmussen managed to hoick out fragments that encode some fifty-five specific known proteins that endow contemporary Yersinia with such lethal virulence. All were present except one, called Yersinia murine toxin (Ymt)—which is the one that allows the bacteria to successfully live inside the flea's gut, by protecting it from being digested. It's Ymt that makes the flea the carrier, but it wasn't there 5,000 years ago. They calculate that it was in place by 1000 BCE, and probably was acquired from another bacteria species donating it. Bacteria can do that; it's called lateral gene transfer, and is an essential part of bacterial evolution. They extend a spear, in Latin a pilus, and short bits of DNA can float across and get incorporated into another bacterial species' genome. We can spot how this has happened in the past due to the presence of flanking bits of DNA that denote a jump from another species.

* There are many ways to do this statistical trick, with different power and reliabilities, but Rasmussen used one called a Bayesian Markov Chain Monte Carlo in software called BEAST 2. We may not be very good at naming human genes, but we make up for it in analytical algorithms.

So, Yersinia in those Bronze Age people was not there by dint of fleas. Without that, and with a slight shift in some of the other genes necessary for delivering full-blown plague, it is likely that infections and outbreaks were not bubonic—the most lethal form—but pneumonic and septicemic. The population of Europe has fluctuated since the introduction of farming over 8,000 years ago, in a sort of boom-and-bust cycle. One of those busts occurred at the time when the fourth millennium BCE was turning into the third, and now we know that plague pathogens were present. It is speculation, but not unreasonable, to suggest that plague had a role in the shaping of this period of Europe's evolution, especially given the annihilations that would follow in recorded history.

Plague is a specter that has been with us for much longer than we can remember. It is the most deadly bacteria known to history, and still haunts us today. These strains are effectively identical to the ones found in the mass graves of east London, but the severity of infection is less. This isn't because of the DNA of Yersinia, and not just because of readily available treatments of modern scientific medicine. We have evolved to cope better with this killer—genetically, socially, and culturally. Pandemics are always just around the corner, but it's now unlikely that plague will devastate the lands as it did repeatedly in history. It is still very present on Earth, and will probably always be so, but we have taken much of the sting out of its tail.

The irony is that *Yersinia pestis* doesn't care about us at all: Its life cycle is primarily enabled by hopping in and out of small mammals that it doesn't particularly hurt, via its preferred host, the flea. We do get fleas and nits, and crabs, but not as much as our more hirsute mammalian distant cousins. The symptoms, the virulence, the catastrophic wave after wave of merciless death that it has handed out indiscriminately to us over five millennia have shaped Europe by crippling armies and toppling empires, and yet it's just a side effect of the blind indifference of natural selection. We brought it on ourselves, by farming and trade and prosperity: Where there is grain, there are rats, and where there are rats there are fleas. Evolution is an arms race, a permanently fluctuating ping-pong between genes in the hosts that carry them and the organisms they feed on, and the spiraling cycles in the struggle for existence. Yersinia found fleas and made them its slave; the flea likes the rat, the rat skulks omnipresent in our shadows. And in the belly of fleas over 5,000 years, with nothing but a simple will to survive via that ancient and universal method of natural selection, *Yersinia pestis* has mutated its genes to do its bidding and, in turn, mutated our own genes in response. DNA sculpts the history of Europe, and the world.

The slow creep out of Africa

The tales of the European adventures are all modern compared to the slow creep from the motherland in eastern Africa. Although still difficult to imagine, that journey played out over centuries or millennia. The nearer we get to the present, the more the increments decrease, and the more graspable human evolution becomes.

It's worth taking a moment to examine the timescales. Out of Africa is a theory that is supported by reams of data and is broadly incontestable today. But what it means when we think of the movement of people over hundreds of generations and thousands of years is deceptively simplistic. Perhaps it is the word *migration* that is so misleading. It means something so specific in modern history, and currently is a hotly debated topic. Migration over an evolutionary timescale is an entirely different phenomenon. A thousand miles over a thousand years is barely movement at all, and the archaeological remains of those movements are scattered wide and far. But that time and those years equally provide the necessary basis for seeing the movement of genes, and how they spread through populations over generational time, of how they are selected by the subtle, largely invisible forces of evolution, and equally how they are deselected. In some senses, archaeology was waiting for genetics to come along to provide a measure that was dependent on unimaginable oceans of time in order for it to make sense.

It's worth stressing again that although getting DNA out of ancient bones is easier now than it ever has been, it's still incredibly difficult. The aim is to get uncontaminated genome-wide samples—not just fragments, but large chunks from all the twenty-three chromosomes. In fact, in most fossils from which DNA can be retrieved, only 1 percent is of the organism itself, the rest being contaminating bacteria that have been feeding off the remains for thousands if not tens of thousands of years. It's in these big pieces of DNA that we can really make sophisticated comparisons and claims about the movement of people in history. There's not exactly an abundance, and although DNA from ancient Europeans is now readily available, there's virtually none from Africans. And the reason is largely to do with heat. DNA kept in a bone in a cold dry-ish cave is much more hardy than DNA stored in a hot humid cave. Although Africa, the nursery of humankind, has given up her bones in abundance, she will not yield the DNA of her children as easily.* But the

* In autumn 2015, a tremendous paper was published in the journal *Science*, in which the first ancient African genome was retrieved from a man who had died and was buried in a cave in the highlands of Ethiopia. The Mota cave is wide and leafy these days, with a pretty sheet of waterfall running for part of the year over the entrance, and into the river the cave overlooks. The man was given the name of his burial site: Mota was a dark-skinned man with dark eyes, lived around 4,500 years ago, died in unsuspicious circumstances, and was buried face up, lying straight with his hands tucked under his chin. His genome was sequenced by a team led by Andrea Manica at Cambridge University, and the paper described revelations, including that there had been a European backflow of anatomically modern humans that had spread and penetrated the genomes of people all over Africa. This was news. It was also news in January

richness of Neolithic DNA, coupled with the genes of modern Europeans, and all the bountiful archaeology that litters European caves and the homes of our ancestors is reconstructing the history of Europe in ever-increasing detail.

All these results are subject to change, or at least refinement, as a sophisticated picture of the movement of people over the world develops, fed by archaeology, language, culture, and now DNA. It is going to be much more complicated than those *Dad's Army* arrows pushing over continental Europe. We can see now where the genes came from, and how culture, specifically farming culture, changed our DNA permanently. Those ghosts in our genomes were just waiting to be discovered, so they could show us how they lived their lives.

And as we progress through time, and from deep history into the last millennium, we enter a time of kings and queens whose lives and lineages are documented in an unprecedented way. The Swedish Motala clan and Loschbour in Luxembourg, and all of those we have resurrected, were just normal people, and only because DNA is the ultimate leveler is their role in the history of humans now being told. Almost everyone who ever lived will be forgotten. But as Europe—and indeed the rest of the world—evolved, tribes grew into fiefdoms, into states, and eventually into empires. People came from the east, and from Africa. Rome petered out, and upon that rock Christianity grew. And while those empires of Europe rose and fell, they were largely oblivious to the existence of two immense continents thousands of miles to the west. The

---

2016, when Manica and his colleagues announced that they had made a mistake. They had, by accident, omitted a step in the statistical computer analysis, which meant that the extent of the backflow was vastly overestimated. The mistake was identified by colleagues and Manica admirably took full responsibility. The genome data itself remains valid, and we await further analysis of Mota's genome. This is science working, and deserves recognition as such—a self-correcting way of acquiring knowledge and understanding.

Viking Leif Ericson was the first European to set foot on these lands, and he called it Vinland—possible meanings include meadow-land, or a reference to the wine-berries or grapevines that they found there. Centuries later and more than two thousand miles farther south from Ericson's first steps into the Americas, Columbus made landfall in what we call the Caribbean. There they met locals, who became generically known to the Spaniards as *indios*—Indians—because Columbus thought he had reached the East Indies. These vast landmasses span just shy of half the circumference of the earth, and the first peoples had been there for more than 20,000 years. But with the arrival of Europeans, these American lands were all set to change.

3

These American lands

Vikings had come from the east, from Greenland, and with threshing oar, drove their ships to new land. Once there, they began an ephemeral exploration of what they called Markland, Helluland, and Vinland. A thousand years ago, according to the Saga of Erik the Red, they found fields of raisin vines, and forests and bears and polar foxes. But they were also met with slingshots and howls from the fur-clad locals. The Norsemen called them Skraeling.

> They were short men, ill-looking, with their hair in disorderly fashion on their heads; they were large-eyed, and had broad cheeks.

For three years the Vikings rambled on these islands, thought now to be Labrador, Newfoundland, and Baffin Island in Canada. They traded cloth for furs. The Skraeling wanted weapons too, but the Vikings forbade arming them with their swords and spears. Thorfinn Karlsefni and Gudrid Thorbjarnardóttir were among the immigrant party, and around the year 1004 CE they had a son, Snorri Thorfinnsson, born in Vinland, and the first person of European ancestry to enter the world in the Americas.

There were constant skirmishes between the Vikings and the Skraeling, and after a scuffle prompted by a loose rampaging Viking bull, the Norsemen decided that though

> the land might be choice and good, there would be always war and terror overhanging them, from those who dwelt there before them.

And so those famously fearsome warriors withdrew. For five centuries, Europeans would not bother the people of the Americas again.

Columbus crossed the ocean blue in 1492, and reached Central America. A few years later Amerigo Vespucci wrote a letter entitled "Mundus Novus" to his patron Lorenzo Pietro di Medici in Florence. Vespucci's first name would be given to the continents north and south. Though it was *mundus novus* for the Italians, the Bahamas was not a new world to the Taíno, for whom it was simply home. Like many of the tribes of the Americas, the Taíno believed that they had always lived on those lands. In their religious tradition, their ancestors emerged from a sacred cave on what we now call Hispaniola.

At the far end of the American continent lies Tierra del Fuego, the southernmost tip of Chile. There, in 1830, Captain Robert Fitzroy docked an exploration vessel, and as part of hostile negotiations seized three Fuegians, boys named *el'leparu* and *o'run-del'lico*, and a girl named *yok'cushly*. They were given absurd English names—York Minster, Jemmy Button, Fuegia Basket—as part of a bizarre colonial experiment to see if these savages could be "civilized." Fitzroy took them to England (a fourth named Boat Memory was also taken, but died of smallpox after they arrived; his real, Fuegian name is lost). Fitzroy brought them back a year later on the HMS *Beagle*'s second voyage. Alongside the three Fuegians was a twenty-two-year-old Charles Darwin, at the beginning of a lifelong journey that would reshape our understanding of life on Earth, and the position of humankind on it.

Europeans arriving in the New World met people all the way from the frozen north to the frozen south. All had rich and mature cultures and established languages. The Skraeling were probably a people we now call Thule, who were the ancestors of the Inuit in Greenland and Canada and the Iñupiat in Alaska. The Taíno were a people spread across multiple chiefdoms around the Caribbean and Florida. Based on cultural and language similarities, we think that they had probably separated from earlier populations from South American lands, now Guyana and Trinidad. The Spanish brought no women with them in 1492, and raped the Taíno women, resulting in the first generation of "mestizo"—mixed ancestry people. Immediately upon arrival, European alleles began to flow, admixed into the indigenous population, and that process has continued ever since: European DNA is found today throughout the Americas, no matter how remote or isolated a tribe might appear to be. But before Columbus, these continents were already populated. The indigenous people hadn't always been there, nor had they originated there, as some of their traditions state, but they had occupied these American lands for at least 20,000 years.

The first Americans

The Americas, North and South, comprise a people simultaneously young and old. The "young" is especially prominent in a country such as the United States, where the genetic, political, and cultural picture of its people is defined by immigrants and the descendants of slaves. The population of all countries is in constant flux, but the modern United States is different, because of this history of slavery, as well as an ever-growing population of South Asians, East Asians, Middle Easterners, and Latinos, who are themselves of multiple lines of descent from Europe, Mesoamerica, and elsewhere. As a result, its underlying contemporary population structure is only just beginning to emerge (and discussed later in this chapter).

The genetic story of African Americans is many books all on its own. It's hard to overstate how complicated a genetic picture they represent, not least because the slave trade compelled a mass migration that is unusual in terms of distance traveled over a short time period, but also because documented ancestry from Africa is mostly absent. Men and women were seized without regard to their birthplace or heritage. Genetics may yet yield many more precise pathways from the origin of humankind in east Africa to the New World, but for now, that picture remains largely opaque (there is further discussion of slavery in Chapter 5). Here, I will focus on the transition that occurred as the lands we now call the Americas became populated by humans for the very first time. All regions of the world are unique, but the tale of the Americas stands alone in its peculiarity and political turmoil.

It's only because of the presence of Europeans from the fifteenth century onward that we even have terms such as *Indians* or *Native Americans*.* How these people came to be is a subject that is complex and fraught, but it begins in the north. Alaska is separated from Russian land by the Bering Strait. There are islands that punctuate those icy waters, and on a clear day US citizens of Little Diomede can see Russians on Big Diomede, just a little over two miles and one International Date Line away. Between December and June, the water between them freezes solid. From 30,000 years ago until around 11,000 BCE, the earth was subjected to a cold snap that sucked up the sea into glaciers and ice sheets extending from the Poles. This period is known as the Last Glacial Maximum, when

* There is little agreement in the academic literature about naming conventions for the various peoples who lived in the Americas before Europeans arrived. I have used *Inuit* broadly as people historically have used *Eskimo*. I've also generally used *Native American* rather than *Indian*, though some people still do, including some Native Americans themselves. Some people prefer to use *indigenous* and avoid any reference to *the Americas*, which is of course an entity that came about only in the modern era.

the reach of the most recent Ice Age was at its fullest. By drilling mud cores out of the seabed, we can reconstruct a history of the land and the seas, notably by measuring concentrations of oxygen, and looking for pollen, which would have been deposited on dry ground from the flora growing there. We think therefore that sea level was somewhere between 60 and 120 meters lower than today. So it was *terra firma* all the way from Alaska to Russia, and all the way down south to the Aleutians—a crescent chain of volcanic islands that speckle the north Pacific.

The prevailing theory about how the people of the Americas came to those lands is via that bridge. We refer to it as a land bridge, though given its duration and size, it was simply continuous land, thousands of miles from north to south; it's only a bridge if we view it in comparison to today's straits.* The area is called Beringia, and the first people across it the Beringians. These were harsh lands, sparse with shrubs and herbs; to the south, there were boreal woodlands, and where the land met the sea, kelp forests and seals.

Though these were still tough terrains, according to archaeological finds Western Beringians were living near the Yana River in Siberia by 30,000 BCE. There's been plenty of debate over the years as to when exactly people reached the eastern side, and therefore at what point after the seas rose they became isolated as the founding peoples of the Americas. The questions that remain—and there are many—concern whether they came all at once or in dribs and drabs. Sites in the Yukon that straddle the US-Alaskan

* Other ideas have been suggested over time, and some have lingered, despite an absence of supporting evidence. The Solutrean Hypothesis was first suggested in 1998, asserting that Europeans had made their way to the Americas during the Last Glacial Maximum, bringing with them stone technology (notably spearheads) seen in Spain, France, and Portugal. Other suggestions have been made over the years, including that Cherokee were from the Middle East, and some Native Americans were descended from lost tribes of Israel. None of these hold water with the application of the slightest scientific scrutiny, but make for thrilling TV documentaries.

border with Canada give us clues, such as the Bluefish Caves, thirty-three miles southwest of the village of Old Crow. These three grottoes are embedded in a Canadian mountain a short scramble up from the Bluefish River. The mountain terrain at this altitude is mostly balding limestone but for a thinning wig of spindly white and black spruce. Inside, just like the Siberian cave in Denisova (see Chapter 1), there is evidence of tens of thousands of years of habitation—bears in winter, foxes in summer, and for a period during the Ice Age, humans.

The latest radio-dating analysis of the remnants of lives in the Bluefish Caves indicates that people were there 24,000 years ago. Beneath the floor of the caves, hundreds of flakes of rock (with exquisite technical names such as burin spalls and lithic debris) made from the action of knapping have been found. And there is also a multitude of worked bone, from caribou, mammoths, horse, and bison.

These founding peoples* spread over 12,000 years to every corner of the continents and formed the pool from which all Americans would be drawn until 1492. I will focus on North America here, and what we know so far, what we can know through genetics, and why we don't know more. Until Columbus, the Americas were populated by pockets of tribal groups distributed up and

---

* A controversial paper was published in April 2017 that purported to show evidence of humans in the Americas as early as 131,000 years ago. Bones from an extinct elephantine beast called a mastodon had been recovered from a dig in California in the 1990s, but in this new study, the authors were suggesting that marks on the bones could only have been achieved by some handiwork thus far only known to be part of the skill set of the genus *Homo*. It is not clear who these people might have been, nor how they might have gotten to California 100,000 years earlier than any other evidence robustly suggests. The response from the academic community has ranged from skeptical to extremely hostile, and the choice phrase from many has been one made popular by the great American astrophysicist Carl Sagan: "Extraordinary claims require extraordinary evidence." For many, this study has not satisfied the second of those criteria. But it's early days, and for now, the new date should be filed in the drawer marked "Extraordinary claims awaiting confirmation."

down both north and south continents. There are dozens of individual cultures that have been identified by age, location, and specific technologies—and via newer ways of knowing the past, including genetics and linguistics. Scholars have hypothesized various patterns of migration from Beringia into the Americas. Over time, it has been suggested that there were multiple waves, or that a certain people with particular technologies spread from north all the way south.

Both ideas have now fallen from grace. The multiple-waves theory has failed as a model because the linguistic similarities used to show patterns of migration are just not that convincing. And the second theory fails because of timing. Cultures are often named and known by the technology that they left behind. In New Mexico there is a small town called Clovis, population 37,000. In the 1930s, projectile points resembling spearheads and other hunting paraphernalia were found in an archaeological site nearby, dating from around 13,000 years ago. These were knapped on both sides—bifaced with fluted tips—and part of the tool kit used for pursuing large megafauna: horse, tapir, sloth, and a menagerie of now extinct beasts, including mammoths, bisons, elephantine mastodons and gomphotheres, and an unwieldy dromedary called a camelop. It had been thought that it was the inventors of these tools who had been the first people to spread up and down the continents. But there's evidence of humans living in southern Chile 12,500 years ago without Clovis technology. These people are too far away to show a direct link between them and the Clovis in such a way that indicates the Clovis being the aboriginals of South America.

Nor is the journey to the southern continent consistent with the Clovis people being the founding population, either—you can't get there from here. To make it to Chile that long ago, their path south from Beringia via land would've been blocked by impassable glaciers. Instead, the working theory is that people may well have gone south in boats, bouncing all the way down the western coast; this is faster, and a similar consistency of climates and geography along the western seaboard buffers against a constant need to adapt to ever-changing environments.

Today, the emerging theory is that the people up in the Bluefish Caves some 24,000 years ago were the founders, and that they represent a culture that was isolated for thousands of years up in the cold north, incubating a population that would eventually seed everywhere else. This idea has become known as Beringian Standstill. Those founders had split from known populations in Siberian Asia some 40,000 years ago, come across Beringia, and stayed put until around 16,000 years ago. Analysis of the genomes of indigenous people show fifteen founding mitochondrial types not found in Asia. This suggests a time when genetic diversification occurred, an incubation lasting maybe 10,000 years. New gene variants spread across the American lands, but not back into Asia, as the waters had cut them off. Nowadays, we see lower levels of genetic diversity in modern Native Americans—derived from just those original fifteen—than in the rest of the world. Again, this supports the idea of a single, small population seeding the continents, and—unlike in Europe or Asia—these people being cut off, with little admixture from new populations for thousands of years, at least until Columbus.

Genetics arms us with the tools to pick apart this model. Nevertheless, we will always rely on hard physical evidence, the traditional form of archaeology. Clovis culture is seen up and down the contiguous United States. There's also the Folsom tradition, slightly later than Clovis, also first discovered in New Mexico; these spearheads are more leaf shaped, have finely worked edges, and were used between 9500 and 8000 BCE. Plano Points culture came next, as the hunter-gatherer tribes of North America continued to develop more efficient ways to kill big beasts. Much of their prey was *Bison antiquus*, a huge beast that wandered up and down the lands en masse during thousands of years of the Last Glacial Maximum. They are all extinct now, though their descendants have become iconic as the American buffalo *Bison bison*.

These technologies clearly served their purpose well. These hunters were seriously effective; some scholars think that the hunting of *Bison antiquus* drove them to extinction. Certainly their remains litter the landscape; they're the most common skeletal remains dragged out of the La Brea Tar Pits,* bang in the middle of urban Los Angeles, a few blocks from Hollywood, but they can also be found up and down North America. Alas, we have far fewer remains of people, and because of this paucity the bodies of the ancients are somewhat iconic. Only one person has been recovered from La Brea—a twenty-five- to thirty-year-old woman from around 10,000 years ago, found with a domestic dog, and therefore thought to be a ceremonial burial. We have no DNA from her.

Elsewhere, we've been a little luckier. In Montana, twenty miles or so off Highway 90, lies the minuscule conurbation of Wilsall, population 178 as of 2010. Though stacks of material culture in the Clovis tradition have been recovered throughout North America, only one person from this time and culture has risen from his grave. He's acquired the name Anzick-1, and was laid to rest in a rock shelter in what would become—around 12,600 years later—Wilsall. He was a toddler, probably less than two years old, judging

* These are an unrivaled source of archaeological interest. The pits have been oozily generating sticky, heavy bitumen oils for thousands of years and trapping unwitting beasts for all that time. They've been dragged for remains throughout the twentieth century, and all manner of beast recovered: saber-toothed cats, wolves, and bears, oh my! But only one human. Universally referred to as "the La Brea Tar pits," the name translated from Spanish tautologically means "the the tar tar pits."

from the unfused sutures in his skull. He was laid to rest surrounded by at least 100 stone tools, and 15 ivory ones. Some of these were covered in red ochre, and together they suggest Anzick was a very special child who had been ceremonially buried in splendor. Now he's special because we have his complete genome.

And there's the woeful saga of Kennewick Man. While attending a hydroplane race in 1996, two locals of Kennewick, Washington, discovered a broad-faced skull inching its way out of the bank of the Columbia River. Over the weeks and years, more than 350 fragments of bone and teeth were eked out of this 8,500-year-old grave, all belonging to a middle-aged man, maybe in his forties, deliberately buried, with some signs of injuries that had healed over his life—a cracked rib, an incision from a spear, a minor depression fracture on his forehead. There have been intensive studies on the dimensions of his bones, with many possible conclusions about whom on Earth he most resembled. Based on the shadows of musculature on his forearms, he seems to have been right handed. There were academic squabbles about his facial morphology, with some saying it was most similar to Japanese skulls, some arguing for a link with Polynesians, and some asserting he must have been European.

With all the toing and froing about his morphology, DNA should be a rich source of conclusive data for this man. But the political controversies about his body have severely hampered his value to science for twenty years. For Native Americans, he became known as the Ancient One, and five clans, notably the Confederated Tribes of the Colville Reservation, wanted to have him ceremonially reburied under guidelines determined by the Native American Graves Protection and Repatriation Act (NAGPRA), which affords custodial rights to Native American artifacts and bodies found on their lands. Scientists sued the government to prevent his reburial, some claiming that his bones suggested he was

European, and therefore not connected with Native Americans. To add an absurd cherry on top of this already distasteful cake, a Californian pagan group called the Asatru Folk Assembly put in a bid for the body, claiming Kennewick Man might have a Norse tribal identity, and if science could establish that the body was European, then he should be given a ceremony in honor of Odin, ruler of the mythical Asgard, though what that ritual entails is not clear.

His reburial was successfully blocked in 2002, when a judge ruled that his facial bones suggested he was European, and therefore NAGPRA guidelines could not be invoked. The issue was batted back and forth for years, in a manner in which no one came out looking good. Nineteen years after this important body was found, the genome analysis was finally published. Had he been European (or Japanese or Polynesian), it would've been the most revolutionary find in the history of US anthropology, and all textbooks on human migration would have been rewritten. But of course he wasn't. A fragment of material was used to sequence his DNA, and it showed that lo and behold, Kennewick Man—the Ancient One—was closely related to the Anzick baby. And as for the living, he was more closely related to Native Americans than to anyone else on Earth, and within that group, most closely related to the Colville tribes.

Anzick is firm and final proof that North and South America were populated by the same people. Anzick's mitochondrial genome is most similar to people of central and south America today (though it should be noted that so far we have no genomes from contemporary Montana tribes, of which there are several). The genes of the Ancient One most closely resemble those of tribes in the Seattle area today. These similarities do not indicate that either were members of those tribes or people, nor that their genes have not spread throughout the Americas, as we would expect over timescales of thousands of years. What they show is that the population dynamics—how ancient indigenous people relate to contemporary Native Americans—is complex and varies from region to region. No people are completely static, and genes less so. Just as we are discovering in Europe and Asia, the patterns of migration of people over millennia is complex and messy, and those precise clean arrows on maps that show where people came from and went are slowly being replaced by loopy, tangled circuit diagrams that reflect the new ancient genetics, and the prototypical human behaviors of movement and enthusiastic reproduction.

In December 2016, in one of his last acts in office, President Barack Obama signed legislation that allowed Kennewick Man to be reburied as a Native American. Anzick was found on private land, so not subject to NAGPRA rules, but was reburied anyway in 2014 in a ceremony involving a few different tribes. The La Brea Woman was removed from display in the tar-pit museum partially out of fear that she might be claimed by Chumash Indians from Southern California. We sometimes forget that though the data should be pure and straightforward, science is done by people, who are never either.

Other pieces of the jigsaw of the first peoples are slowly beginning to fall into place. At the end of 2016, an earlier admixture was revealed in the Inuit of Greenland. These people live in the cold, and their diet is dominated by seafood. By sampling hundreds of Inuit genomes, an area of DNA had been identified that plays a part in these very Inuit characteristics. One of the genes that sits there, TBX15 (part of a family of genes called transcription factors—more on these in Chapter 6), does a huge range of things in the body by regulating the expression of other genes. The version of TBX15 found in the Inuit appears to have a role in the distribution of fat around the body, and may help maintain warmth by fueling a particular type of fat burning. The latest work, by UC Berkeley's

Rasmus Nielsen, shows that not only had this region been undergoing positive selection in the last few thousand years, but also that this particular variant appears to be a piece of Denisovan DNA. It's virtually absent in Africa, and different from the one we see in Neanderthals. This suggests that the people of Beringia had this Inuit version, acquired from their relations with the Denisovans, and in those cold climes it proved to help hardy fishermen.* We see a similar story in other genes relating to diet. Fatty acid desaturases (FADS) are enzymes that help convert saturated fats found in fish and meat into unsaturated fats. They're a rich source of evolutionary intrigue, too: Many studies have revealed that this cluster of genes shows signs of positive selection in ancient populations all over the world. It's been examined in those early Europeans Loschbour and Stuttgart (see Chapter 2), as well as in Denisovans and Neanderthals, and shows a complex distribution that is not easy to explain. But what is clear is that there have been selective sweeps around these genes in all populations, suggesting their importance in adapting to the foods available to our ancient forebears in their local environment. The Greenland Inuit have versions of the FADS genes that also look like they have been positively selected in Beringia more than 18,000 years ago. This is not wholly unexpected given that FADS appear to evolve in response to local conditions all over the world. But we can use this fact as a tracer for migration in the Americas. By examining the distribution of the Inuit version of the FADS genes up and down the Americas, Spanish gene hunters led by Tábita Hünemeier showed in 2017 that fifty-three indigenous populations in the Americas showed local adaptation in the FADS that were all derived from a founder population in Beringia. This includes a forceful presence in Amazonian tribes, where diet and lifestyle is obviously different from the Inuits'.

---

* Rasmus Nielsen: personal communication.

These gene maps powerfully suggest that that the Beringian Standstill hypothesis is correct: All Native Americans, north and south, have versions of genes relating to diet that are suited to their current environments, but born of an ancient population subject to local adaptation in the frozen north, thousands of years ago.

Just as in the rest of the world, stories are beginning to emerge that explain why people are the way they are, and how they got there. But despite this progress, understanding the genetics and the genetic genealogy of the indigenous people of the Americas, particularly North America, is proving to be far trickier than it should. Anzick and Kennewick Man represent narrow samples—a tantalizing glimpse of the big picture. And politics and history are hampering progress.* The legacy of 500 years of occupation have fostered profound difficulty in understanding how the Americas were first peopled. Two of the doyennes of this field—Connie Mulligan and Emőke Szathmáry—suggest that there is a long cultural tradition that percolates through our attempts to deconstruct the past. Europeans are taught a history of migration from birth, of Greeks and Romans spreading over Europe, conquering lands and interloping afar. Judeo-Christian lore puts people in and out of Africa and Asia, and the silk routes connect Europeans with the east and back again. In Britain we learn of Vikings and Saxons and others coming over from mainland Europe and spreading their seeds, as detailed in the previous chapter. Many European countries have been seafaring nations, exploring and sometimes belligerently building empires, for commerce or to impose a perceived superiority over other people. Even though we Brits state (with some degree of truth) that we haven't been aggressively invaded since 1066, the European story is one in which the movement of people is inherent. Even though we

---

* The book *Native American DNA* by Kim TallBear is the definitive cultural analysis of genetics in relation to the peoples of the Americas, and has influenced my thinking on this complex matter.

have national identities, and pride and traditions that come with that sense of belonging, European culture is imbued with migration.

For Native Americans, this is not their culture. Not all believe they have always been in their lands, nor that they are a static people. But for the most part, the narrative of migration does not threaten European identity in the same way that it might for the people we called the Indians. The scientifically valid notion of the migration of people from Asia into the Americas may challenge Native creation stories. It may also have the effect of conflating early modern migrants from the fifteenth century onward, with those from 24,000 years earlier, with the effect of undermining indigenous claims to land and sovereignty.

This we have seen before in other lands subjected to colonial rule: My father was raised in New Zealand in the 1950s, where he was taught that the indigenous Maori were not the first people, but that they had conquered a more primitive tribe called the Moriori. This was in fact a fabrication perpetrated by British–New Zealander writers in the nineteenth century, who used it as justification for their own colonization: Maori superseded a less advanced society, and now Europeans are merely doing the same. These big cultural differences have been all too infrequently acknowledged in interactions between scientists and Native Americans. Together they form a suite of reasons why there is a residual, background animosity toward science from some Native American communities. There are other reasons that are far more pernicious.

The tale of the Havasupai

Deep among the lakes of the Grand Canyon are the Havasupai. Their name means "people of the blue-green waters," and they've been there for at least 800 years. They're a small tribe, around 650 members today, and they use ladders, horses, and sometimes helicopters to travel in and out—or rather, up and down—the canyon. The tribe is rife with type 2 diabetes, and in 1990, the Havasupai people agreed to provide Arizona State University scientists with DNA from 151 individuals with the understanding that they would seek genetic answers to the puzzle of why diabetes was so common. Written consent was obtained, and blood samples were taken.

An obvious genetic link to diabetes was not found, but the researchers continued to use their DNA to test for schizophrenia and patterns of inbreeding. The data was also passed on to other scientists who were interested in migration and the history of Native Americans. The Havasupai only found this out years later, and eventually sued the university. In 2010, they were awarded $700,000 in compensation. Therese Markow was one of the scientists involved, and insists that consent was on the papers they signed, and that the forms were necessarily simple, as many Havasupai do not have English as a first language, and many did not graduate from high school. Part of the question here is really of the nature of consent. "Informed consent" is the standard phrase used in performing any medical or scientific investigation on volunteers. A properly administered study affords the volunteer full understanding of how their tissue or sample will be analyzed and used. Markow's lawyers argued that the consent specified other uses, as per the phrase to "study the causes of behavioral/medical disorders," but many in the tribe thought that they were only being asked about their endemic diabetes. A blood sample contains an individual's entire genome, and with it, reams of data about that individual, their family, and evolution.

This isn't the first time this has happened. In the 1980s, before the days of easy and cheap genomics, blood samples were taken with consent to analyze the unusually high levels of rheumatic disease in the Nuu-chah-nulth people of the Pacific Northwest of

Canada. The project, led by the late Ryk Ward, then at the University of British Columbia, found no genetic link in their samples, and the project petered out. By the '90s, though, Ward had moved to the University of Utah, and then Oxford in the UK, and the blood samples had been used in anthropological and HIV/AIDS studies around the world, which turned into grants, academic papers, and a PBS/BBC jointly produced documentary.

Ward's conclusions about the origin of the Nuu-chah-nulth were not correct, giving them an earlier origin than via Beringia some 20,000 years ago. The use of the samples for historical migration indicated that the origins of the Havasupai were from ancient ancestors in Siberia, which is in accordance with our understanding of human history by all scientific and archaeological methods. But it is in opposition to the Havasupai religious belief that they were created *in situ* in the Grand Canyon. Though nonscientific, it is perfectly within their rights to preclude investigations that contradict their stories, and those rights appear to have been violated. Havasupai vice chairman Edmond Tilousi told *The New York Times* in 2010 that "coming from the canyon . . . is the basis of our sovereign rights."

Sovereignty and membership of a tribe is a complex and hard-won thing. It includes a concept called "Blood Quantum," which is effectively the proportion of one's ancestors who are already members of a tribe. It's an invention of European Americans in the nineteenth century, and though most tribes had their own criteria for tribal membership, most eventually adopted Blood Quantum as part of the qualification for tribal status.

DNA is not part of that mix. With our current knowledge of the genomics of Native Americans, there is no possibility of DNA being anywhere near a useful tool in ascribing tribal status to people. Furthermore, given our understanding of ancestry and family trees (discussed in more detail in the following chapter), I have profound doubts that DNA could *ever* be used to determine tribal membership. While mtDNA (which is passed down from mothers to children) and the Y chromosome (passed from fathers to sons) have both proved profoundly useful in determining the deep ancestral trajectory of the first peoples of the Americas into the present, these two chromosomes represent a tiny proportion of the total amount of DNA that an individual bears. The rest, the autosomes, comes from all of one's ancestors.

In the previous chapter, I talked about how some genetic genealogy companies will sell you kits that claim to grant you membership to historical peoples, albeit ill-defined, highly romanticized versions of ancient Europeans. This type of genetic astrology, though unscientific and distasteful to my palate, is really just a bit of meaningless fantasy; its real damage is that it undermines scientific literacy in the general public. Over centuries, people are too mobile to have remained genetically isolated for any significant length of time. Tribes are known to have mixed before and after colonialism, which should be enough to indicate that some notion of tribal purity is at best imagined. Of the genetic markers that have been shown to exist in individual tribes so far, none is exclusive. Some tribes have begun to use DNA as a test to verify immediate family, such as in paternity cases, and this can be useful as part of qualification for tribal status. But on its own, a DNA test cannot place someone in a specific tribe.

That hasn't stopped the emergence of some companies in the United States that sell kits that claim to use DNA to ascribe tribal membership. Accu-Metrics is one such company. On their web page, they state that there are "562 recognized tribes in the U.S.A., plus at least 50 others in Canada, divided into First Nation, Inuit, and Metis." For $125 they claim that they "*can determine if you belong to one of these groups.*"

The list is comprehensive, from Abenaki to Zuni. Accu-Metrics is not the only company: DNA Consultants sells a Cherokee test for $99. They claim a database of sixty-two Cherokee genomes, and if your match is high enough, you qualify for a certificate (for an extra $25). If it does not match their Cherokee database but does indicate some Native American DNA, you also get a certificate (also costing an extra $25). As with all genetic genealogy tests, these may show that you share parts of your DNA with Native Americans today. If you are trying to establish immediate family relations, and they are enrolled in a tribe, then a DNA test may help your own enrollment. But otherwise, there is no biological test that alone can demonstrate tribal membership.

Still, some people are convinced otherwise. Carol Reynolds Boyce has long believed she was a member of the Beothuk, a Newfoundland tribe whose last known member died in 1829. They were driven to extinction by a combination of exposure to smallpox, tuberculosis, and persecution. But we do have some of their remains in DNA: Small fragments of genome are known from two Beothuk—Demasduit and her husband, Nonosabasut, who also died in the 1820s—extracted from their teeth. These fragments reveal two mtDNA types that are typical of Native Americans, notably another Newfoundland tribe called the Mi'kmaq. These two tribes were shown to not be particularly closely related in 2011, which is not really the point. The problem is twofold: First, the fragments of Beothuk DNA from Demasduit and Nonosabasut are not nearly enough to constitute a valuable database from which to make a comparison with DNA from anyone else. The second problem is more general: DNA is not unique to any one tribe.

Carol Reynolds Boyce received her test results from Accu-Metrics in October 2016, which to her mind confirmed her own beliefs, passed down orally by relatives: She is descended from Beothuk. She asserts that this DNA test is the basis of the creation of what she is calling the Beothuk First Nation, and has written to the Canadian Prime Minister Justin Trudeau to alert the government to her claims. In a report in the Newfoundland newspaper the *Telegram* in January 2017, she clarified her intentions: "Let me make it perfectly clear. Beothuk First Nation is in progress to get federal recognition."

Accu-Metrics Lab manager Kyle Tsui told the *Telegram* that the "test that we do is not a legal test. . . . You can't use it as any evidence whatsoever, it's just for informational purposes. You can't take it to court and so forth." In January 2017, following media interest, Accu-Metrics decided to withdraw the Beothuk test. At the time of writing (June 2017), Accu-Metrics' website still says this: "All tests can be done for legal purposes" and "The results of this scientific test can be used to receive a status card or tribal enrollment."

Maybe Carol Reynolds Boyce can establish a Beothuk First Nation. But the idea that tribal status is encoded in DNA is both simplistic and wrong. Many tribespeople have nonnative parents and still retain a sense of being bound to the tribe and the land they hold sacred. In Massachusetts, members of the Seaconke Wampanoag tribe identified European and African heritage in their DNA, due to hundreds of years of interbreeding with New World settlers. Attempting to conflate tribal status with DNA denies the cultural affinity that people have with their tribes. It suggests a kind of purity that genetics cannot support, a type of essentialism that resembles scientific racism.

There is a deep history of real, murderous, nonscientific racism toward American Indians by Europeans. Native Americans have been slaughtered, have had their lands and possessions taken, and have been persecuted for hundreds of years. President Andrew Jackson enacted the Indian Removal Act in 1830, which gave the federal government the right to negotiate relocation of certain southern tribes of the Cherokee Nation. Though discussions were technically voluntary, enormous pressure was put on the tribal chiefs to relocate away from their own lands. This preceded the forced removal of other tribes, known as the Trail of Tears. Following the discovery of gold in Cherokee territory, more than 16,000 were forcibly moved, and though the numbers are uncertain, many thousands of Native Americans died en route.

The stigmatization and persecution of Native Americans continued well into the twentieth century, too. There are too many examples to name, and they span the whole of North America. Because of their proximity to the Japanese, 881 Aleuts of the Alaskan island chain were interned during the Second World War following the attack on Pearl Harbor, their houses burned by US troops to prevent the Japanese from using them. The Aleuts were housed in conditions far worse than the 700 Nazis who were captured in North Africa and imprisoned a few hundred miles away in Alaska. In Chapter 5, on the relationship between genetics and race, we'll see how thirty-one of the United States enacted eugenics policies, and how Native American populations were disproportionately affected, with thousands being forcibly sterilized as recently as the 1970s. The stigma of a genetic predisposition to alcoholism remains among Native Americans to this day, despite the fact that it is a claim not rooted in fact.

The specious belief that DNA can bestow tribal identity, as sold by companies such as Accu-Metrics, can only foment further animosity—and suspicion—toward scientists. If a tribal identity could be shown by DNA (which it can't), then perhaps reparation rights afforded to tribes in recent years might be invalid in the territories to which they were moved during the nineteenth century. Many tribes are effective sovereign nations and therefore not necessarily bound by the laws of the state in which they live.

When coupled with cases such as that of the Havasupai, and centuries of racism, the relationship between Native Americans and geneticists is not healthy. After the legal battles over the remains of Kennewick Man were settled, and it was accepted that he was not of European descent, the tribes were invited to join in the subsequent studies. Out of five, only the Colville Tribes did. Their representative, James Boyd, told *The New York Times* in 2015, "We were hesitant. Science hasn't been good to us."

Following the award of damages to the Havasupai, Ron Whitener, a Native American ethics expert at the University of Washington School of Law in Seattle said that it "puts every tribe in the US on notice regarding genetics research." This is a twisted, vicious circle that nurtures a gap in our knowledge. In an article in the *American Journal of Medical Genetics*, Whitener added, "Other tribes see this as a disrespect to the sovereignty of tribal government. . . . As Native Americans, this case informs us that tribes need to provide effective oversight of research, especially genetic studies."

Perhaps the lesson from the tale of the Havasupai—and other examples of the sometimes troubling interactions between modern science and Native Americans—is not in the details of whether consent was explicit or explicitly informed, or the legal ping-pong over Kennewick Man, but that the engagement between the geneticists and the tribe was neither respectful nor courteous. It was also scientifically shortsighted. With culturally inept engagement, and documented, subsequent animosity from tribes to science, scientists can't access the DNA that would answer so many burning questions that are now easy to address.

However broad consent was, understanding of why the Havasupai were disturbed by the further use of their blood does not appear to have been sought. It is my suspicion that this was clumsy rather than willful exploitation. But maybe it demonstrates a lack of grace in understanding that people have different views about their cultures and traditions. I don't pretend to understand Native American culture. It is not one thing; they are not a monolithic people. I can read about it, and talk to Native Americans, but a lived life is shaped by history and culture. For many, the land is sacred. For many, DNA is viewed as part of the body, which of course it is, and this makes it sacred, too. It ties a person's identity to the land and to their ancestors. Genetics has become a historical subject, but as argued elsewhere in this book, DNA is a historical source that is complementary to others, and geneticists should avoid the temptation to assert its supremacy over other ways of knowing the past and of knowing a people. Human genetics is the study of humans, who should not be viewed simply as specimens, but as people.

The scientific—if naïve—desire to know the Americas is itself problematic, revealing a form of colonialism that persists within science. Whether it is malicious or simply flat-footed is difficult to ascertain, and probably deserves scrutiny on a case-by-case basis. Data is supreme in genetics, and data is what we crave.

But *we* are the data, and people are not there for the benefit of others, regardless of how noble one's scientific aims are. To deepen our understanding of how we came to be and who we are, scientists must do better, and invite people whose genes provide answers to not only volunteer their data, but to participate, to own their individual stories, and to be part of that journey of discovery.

This is beginning to change. A new model of engagement with the first people of the Americas is emerging, albeit at a glacial pace. The American Society for Human Genetics meeting is the annual who's who in genetics, and has been for many years, where all of the newest and biggest ideas in the study of human biology are discussed. In October 2016 they met in Vancouver, and it was hosted by the Squamish Nation, a First Nations people based in British Colombia. They greeted the delegates with song, and passed the talking stick to the president for the proceedings to begin. The relationship between science and indigenous people has been one characterized by a range of behaviors from outright exploitation to casual insensitivity to tokenism and lip service. Perhaps this time is coming to an end and we might foster a relationship based on trust, genuine engagement, and mutual respect, so that we might work together and build the capacity for tribes to lead their own research into the histories of these nations.

The New World

Today, we live in this age of ancient genetics where we can wheedle out the DNA of peoples long dead and reconstruct past lives. The genomics of Europe is building up the most detailed prehistory we have ever known. Asia is following suit. Africa will always be more difficult because the heat provides a less-than-ideal protector of DNA, fragile and capricious as it can be in the bones of the dead.

Though the terms *Native American* and *Indian* are relative, the United States is a nation of immigrants and descendants of slaves who have overwhelmed the indigenous population. Less than 2 percent of the current population defines itself as Native American, which means that 98 percent of Americans are unable to trace their roots, genetic or otherwise, beyond 500 years on American soil. That is, however, plenty of time for populations to come and breed and mix and lay down patterns of ancestry that can be enlightened with living DNA as our historical text. A comprehensive genetic picture of the people of postcolonial North America was revealed at the beginning of 2017, drawn from data submitted by paying customers to the genealogy company AncestryDNA. The genomes of more than 770,000 people born in the USA were filtered for markers of ancestry, and revealed a picture of mish-mash, as you might expect from a country of immigrants. Nevertheless, just as in the People of the British Isles project (described in Chapter 2), genetic clusters of specific European countries are seen. Paying customers supply spit harboring their genomes, alongside whatever genealogical data they have. By aligning these as carefully as possible, a map of post-Columbus America can be summoned with clusters of common ancestry, such as Finnish and Swedish in the Midwest, and Acadians—French-speaking Canadians from the Atlantic seaboard—clustering way down in Louisiana, close to New Orleans, where the word *Acadian* has mutated into *Cajun*. Here, genetics recapitulates history, as we know the Acadians were forcibly expelled by the British in the eighteenth century, and many eventually settled in Louisiana, then under Spanish control.

In trying to do something similar with African Americans, we immediately stumble. Most black people in the United States cannot trace their genealogy with much precision because of the legacy of slavery. Their ancestors were seized from West Africa, leaving little or no record of where they were born. In 2014, the genetic genealogy company 23andMe published their version of the population structure of the United States. In their portrait we see a similar pattern of European admixture, and some insights into the history of the postcolonial United States. The Emancipation Proclamation—a federal mandate to change the legal status of slaves to free—was issued by President Lincoln in 1863, though the effects were not necessarily immediate. In the genomic data, there's admixture between European DNA and African that begins in earnest around six generations ago, roughly in the mid-nineteenth century. Within these samples we see more male European DNA and female African, measured by Y chromosome and mitochondrial DNA, suggesting male Europeans had sex with female slaves. Genetics makes no comment on the nature of these relations.

The greatest horseman in the world

Just as so many Europeans want to discover that DNA will unveil Viking blood, the discovery of Native American ancestry offers cachet and kudos. But matters are not so simple.

This is a story with no end, but instead I offer you a coda—an intriguing but cautionary tale of the messy business of genealogy. Genetic ancestry tests will not tell you what tribe you are a member of. But we do know enough about the genomes of indigenous Americans for it to show up in a blurry, highly generalized way.

The company 23andMe color-codes its maps, which show where on Earth DNA similar to yours can be found today. My map is almost entirely predictable. 50.1 percent is yellow, which is their code for southern Asia, meaning India. Forty-nine percent is hues of blue, meaning European. My mother is Indian, though she was born in South America (we don't know her deeper family history; her ancestors were dispatched to Guyana as indentured laborers, but records were lost in a fire). My father is British, born in Yorkshire, where I still have relatives, and they have drawn out much of our family tree back a few centuries, which meanders around the UK, London, bits of Ireland, but mostly in the northeast of England and into Scotland, as befits the Clan Rutherford.

Some of the missing 0.9 percent is marked red, the color code for the Americas. This fraction of my genome is identified as "Native American." This was not expected.

My family's tree was compiled by my father's cousin, a keen amateur genealogist. In Yorkshire, Kevin O'Byrne has been using the traditional forms of family detective work to make his way up our family tree—births, deaths, and marriages, church records, military service records, business receipts, and so on. He has been compiling a family tree for the branches we can trace back into the eighteenth century and beyond. At the time of writing, he had found one of my great-great-great-grandfathers, a man with the splendid moniker Lycurgus Handy. We know little of Lycurgus, other than that his life did not match the grandeur of his name. He died destitute and diseased in the poorhouse in Hackney, east London, in 1920, a stone's throw from where my own son was delivered by me in haste in our bedroom. I carved notification of this event into wet concrete when the foundations were later reset, for some future historian to ponder. Kevin had to look harder, though, as genealogical records are rarely written in wet concrete. Nevertheless, he kept digging and found that Lycurgus' grandfather was one Benjamin Handy. Ben had a professional nickname: He was known as "the greatest horseman in the world."

Ben Handy was proprietor and star performer in Handy's Traveling Circus, a popular touring act that made its business by bringing wonder all over Britain and Ireland in the eighteenth century. He was married to another performer, Mary Huntley. On their wedding certificate, dated 1818 in St Paul's in Covent Garden, London, it describes her as "savage." We believe (so far, as this investigation is currently active) that Mary was the daughter of Neil Huntley, one of two "chiefs" from the American tribe called the Catawba, who were brought from America to join the show for their exotic native horse skills. The Catawba are a federally recognized tribe whose land is on the border of South Carolina. Could this be the source of the flash of Native American DNA in my genome?

Elsewhere in this book, I express skepticism about what genomes can tell us about families and our pasts. Generally, with commercial genetic genealogy testing companies, the data is good, but the interpretation of that data can be extremely creative. If real, then this proportion of Native American DNA will tell me nothing about myself, and it's less than the amount of Neanderthal DNA I bear. Still, it's quite exciting, because it has the beginnings of an interesting tale.

But here's the kicker: It could be just noise. The genetics of Native Americans are poorly understood, and there are no specific tribal markers in DNA. These commercial databases are primarily populated with the data of paying customers. Contribution to the 23andMe database is low for these diverse peoples—1 percent in the aforementioned study from 2015. So it could be that my red flash is a very small percentage that is misidentified due to a small sample. The raw data is not easily accessible, so even for a geneticist, it isn't easy to check.

Narrative satisfaction is what we crave, and as mentioned elsewhere, genetics rarely delivers. This tale is no different. I include it here because it is interesting and exciting, but also because it highlights the intrigue that family genealogy invites, only to be matched with the cold reality of the messiness of family histories, and genetics. It is likely that those very low numbers of genetic markers that look like this and are found in the Americas are simply noise. Other members of my family have done their own ancestry analyzed by the same method, but not by the same company, and have so far drawn a blank. That may not mean mine is incorrect, but it doesn't help it to be true. If it is correct, if I am descended from a Catawba chief, it doesn't make me a Native American, and it doesn't afford me any of the cultural or political valences that come with membership of a tribe. Neil Huntley, if indeed he was Catawba, is one of 128 male ancestors from that tier of my family tree.

Even if the genetics and the genealogy are correct, it is equally possible—and indeed in my opinion more likely—that the sliver of supposedly Native American DNA in my genome is merely noise in a file comprising hundreds of thousands of data points. Even if the family tree is correct, it is quite possible that I harbor literally no DNA from Neil Huntley and the Catawba people, as it has been diluted through generations. It's exciting to discover circus performers in one's family tree, but I must be very careful not to fall into the trap I am decrying throughout this book, that DNA has some power to determine identity.

The truth is that it means next to nothing. Ancestry is a matted web. There would be thousands, probably millions, of Europeans with similar claims, if their family trees could be drawn in such detail.

The cachet of having interesting ancestors is compelling, whether they are Catawba, or Beothuk, or Vikings, or Aleut, or Skraeling. But do not be disheartened if you can't find warriors, hunters, or royalty in your tree. In the next chapter, we will find out that we are all descended from kings.

4

When we were kings

i: The king lives on

Charlemagne, Carolingian King of the Franks, Holy Roman Emperor, the great European conciliator; your ancestor. I am making an assumption that you are broadly of European descent, which is not statistically unreasonable but certainly not definitive. If you're not, be patient, and we'll come to your own very regal ancestry soon enough.

Along with Alexander and Alfred, Charlemagne is one of a handful of kings who gets awarded the post-nominal accolade "the Great." His early life remains mysterious and the stories are assembled from various sources, but it seems he was born around 742 CE, just at the time when the Plague of Justinian was dispatching millions at the eastern edge of the moribund Roman Empire. The precise place of his birth is also unknown, but it's likely to be in a town such as Aachen, now in contemporary Germany, or Liège in Belgium. Even Einhard, his dedicated servant and biographer, wouldn't get drawn into the specifics of Charlemagne's early life in his fawning magnum opus, *The Life of Charles the Great*. The very fact that this account exists—probably the first biography of a European ruler—is testament to how important he was (or at least was seen to be). In many European languages, the word *king* is itself derived from Charlemagne's name.

He was the son of Pippin the Short,* an aggressive ruler of France who expanded the Frankish kingdom until his death during the return journey from a campaign against the persistently rebellious realm of Aquitaine in 768. Charlemagne stepped up as his successor, and continued the expansion with aplomb. He battled the Saxons to the northeast, the Lombards in Italy, and Muslims in Spain. He capitalized on his father's good political relations with the Vatican, and in 800 was crowned the first Holy Roman Emperor by Pope Leo III in St Peter's Basilica, an event so momentous that Charlemagne marked it by giving Leo one of the great medieval relics as a thank you—the Holy Prepuce, better known as Jesus' foreskin.†

A fecund ruler, Charlemagne sired at least eighteen children by motley wives and concubines, including nine by his second wife, Hildegard of Vinzgau. These kin included Charles the Younger, Pippin the Hunchback, Drogo of Metz, Hruodrud, Ruodhaid, Adalheid, Hludowic, and not forgetting Hugh, and he consolidated his reign by installing many of his sons in positions of power across the expanding empire. Royal lineages are historically the only ones to get documented well until the modern era, and Charlemagne's lineage is bountiful. We can trace a path directly from his fruitful loin: It begins with his son Louis the Pious, via Lothar, Bertha, Willa, Rosele, eight men called Baldwin, and so on through the ages until it reaches the twenty-first century in a Dutch family

---

* who was not particularly short: this is probably a mistranslation of Pepin le Bref, or "the younger." Charlemagne himself was tall; his tibia was recently measured to be seventeen inches, which means he was at least five feet ten inches, and maybe as much as six four. His male contemporaries were on average five six.

† This was the first documented foreskin of Christ, though over the years there have been at least eighteen, all over the world, in all sorts of ceremonies. One seventeenth-century scholar, the Vatican librarian Leo Allatius, suggested in a treatise that Jesus' foreskin ascended into the heavens to become the rings of Saturn. Support for this argument is stretched thin.

called the Backer-Dirks, whose family tree all the way back to the king is publicly available online.

This pedigree, by gorgeous chance, also contains Joachim Neumann, a seventeenth-century German Protestant preacher, who sought peace and meditation away from the political machinations and church hullabaloo of Dusseldorf in a small cave near the river Dussel. He had changed his name from Neumann to a Greek version with the same meaning: new man. He wasn't the only new man in that cave. This was the location of the very first new human to be identified, a century later, in the valley of Joachim Neander—Neanderthal man.

What a lineage to behold! It comes as no great surprise that in the world of amateur genealogy, being descended from imperial royalty is considered of high cachet. In fact, descent from anyone actually named from history brings prestige, as the vast majority of humans have drifted into and out of existence leaving little or no historical footprint that shows they ever drew breath. To be drawn from the bloodline of a king, and not just any old Holy Roman Emperor, but the very first, must be momentous.

Christopher Lee—the great actor who among his roles counts Dracula; Tolkien's Saruman the White; the Man with the Golden Gun himself, Scaramanga; the fallen Jedi Count Dooku; and *The Wicker Man*'s Lord Summerisle—claimed direct ancestry to King Charlemagne via the ancient house of his mother, Countess Estelle Marie (née Carandini di Sarzano):

> The Carandini family is one of the oldest in Europe and traces itself back to the first century AD. It is believed to have been connected with the Emperor Charlemagne, and as such was granted the right to bear the coat of arms of the Holy Roman Empire by Emperor Frederick Barbarossa.*

* As reported by University College Dublin on awarding Christopher Lee Honorary Life Membership of their Law Society in 2011.

Maybe it was to enhance his august yet sinister screen image as having played some of the wickedest characters in film history. Most people don't have a coat of arms, but I can say with absolute confidence that if you're vaguely of European extraction, just like cinema's greatest Prince of Darkness, you are descended from Charlemagne too. Hail to the king!

We are all special, which also means that none of us is. This is merely a numbers game. You have two parents, four grandparents, eight great-grandparents, and so on. Each generation back the number of ancestors you have doubles. But this ancestral expansion is not borne back ceaselessly into the past. If it were, your family tree when Charlemagne was Le Grand Fromage would harbor around 137,438,953,472 individuals on it—more people than were alive then, now, or in total. What this means is that pedigrees begin to fold in on themselves a few generations back, and become less arboreal, and more a mesh or weblike. You can be, and in fact are, descended from the same individual many times over. Your great-great-great-great-great-grandmother might hold that position in your family tree twice, or many times, as her lines of descent branch out from her, but collapse onto you. The further back through time we go, the more these lines will coalesce on fewer individuals. *Pedigree* is a word derived from the middle French phrase *pied de grue*—the crane's foot—as the digits and hallux spread from a single joint at the bottom of the tibia, roughly equivalent to our ankle. This branching describes one or a few generations of a family tree, but it's wholly inaccurate as we climb upward into the past. Rather, each person can act as a node into whom the genetic past flows, and from whom the future spills out, if indeed they left descendants at all.

This I find relatively easy to digest. The simple logic is that there are more living people on Earth now than at any single moment in the past, which means that many fewer people act as multiple ancestors of people alive today. But how can we say with utter confidence that any individual European is, like Christopher Lee, directly descended from the great European conciliator?

The answer came before high-powered DNA sequencing and ancient genetic analysis. Instead it comes from mathematics. Joseph Chang is a statistician from Yale University and wished to analyze our ancestry not with genetics or family trees, but just with numbers. By asking how recently the people of Europe would have a common ancestor, he constructed a mathematical model that incorporated the number of ancestors an individual is presumed to have had (each with two parents), and given the current population size, the point at which all those possible lines of ascent up the family trees would cross. The answer was merely 600 years ago. Sometime at the end of the thirteenth century lived a man or woman from whom all Europeans could trace ancestry, if records permitted (which they don't). If this sounds unlikely or weird, remember that this individual is one of thousands of lines of descent that you and everyone else has at this moment in time, and whoever this unknown individual was, they represent a tiny proportion of your total familial webbed pedigree. But if we could document the total family tree of everyone alive back through 600 years, among the impenetrable mess, everyone alive of European descent would be able to select a line that would cross everyone else's around the time of Richard II.

Chang's calculations get even weirder if you go back a few more centuries. A thousand years in the past, the numbers say something very clear, and a bit disorienting. One fifth of people alive a millennium ago in Europe are the ancestors of no one alive today. Their lines of descent petered out at some point, when they or one of their progeny did not leave any of their own. Conversely, the remaining 80 percent are the ancestor of everyone living today. All lines of ancestry coalesce on every individual in the tenth century.

One way to think of it is to accept that everyone of European descent should have billions of ancestors at a time in the tenth century, but there weren't billions of people around then, so try to cram them into the number of people that actually were. The math that falls out of that apparent impasse is that all of the billions of lines of ancestry have coalesced into not just a small number of people, but effectively literally everyone who was alive at that time. So, by inference, if Charlemagne was alive in the ninth century, which we know he was, and he left descendants who are alive today, which we also know is true, then he is the ancestor of everyone of European descent alive in Europe today.

It's not even relevant that he had eighteen children, a decent brood for any era. If he'd had one child who lived and whose family propagated through the ages until now, the story would be the same. The fact that he had eighteen increases the chances of his being in the 80 percent rather than the 20 percent who left no twenty-first century descendants, but most of his contemporaries, to whom you are all also directly related, will have had fewer than eighteen kids, and some only one, and yet they are all also in your family tree, unequivocally, definitely and assuredly.

At least that's the theory. With the advent of easy and cheap DNA sequencing came the possibility of testing this math. DNA is the bearer of biological ancestry, and you get all of your DNA from your two parents, pretty much a 50:50 split. They in turn got all of their DNA from their parents, so one quarter of your DNA is the same as a quarter of each of your grandparents. If you have a cousin, then you share around an eighth of your DNA, as you have a pair of grandparents in common. These shared bits of DNA are not the same sections though. And it doesn't keep halving perfectly as you meander up through your family tree. Remember that DNA gets shuffled when a sperm or egg is made, and every single shuffle is different, but it's quite clumsy shuffling. In the newly shuffled deck, that is, your own personal genome, big chunks of it are the same as your father or mother. The more closely related two people are, the more DNA they will share in big chunks. This is why identical twins are identical (all the chunks are the same), and why siblings and parents look similar (half of their DNA is the same as each other). In genetics, we call these sections of DNA identical by descent (IBD), and they are very useful for measuring the relatedness of two individuals.

In 2013, geneticists Peter Ralph and Graham Coop showed that DNA says exactly the same thing as Chang's mathematical ancestry: Our family trees are not trees at all, but entangled meshes. They looked for lengths of identical by descent DNA in 2,257 people from around Europe (to mitigate the influence of recent migration, all the subjects selected had four grandparents from the same region or country). By measuring the lengths of the shared DNA, they could estimate how long ago that deck got shuffled, and therefore how related any two people are. Computing and DNA have empowered this field, and this is shown in their dataset and the number crunching that follows.

Joseph Chang's mathematical calculation didn't account for something very obvious, which is that we don't mate randomly. We typically marry within socioeconomic groups, within small geographical areas, within shared languages. But with Coop and Ralph's genetic analysis, it didn't seem to matter that much. Ancestry is such that genes can spread very quickly over generations. It might seem that a remote tribe would have been isolated from others for centuries in, for example, the Amazon. But no one is isolated indefinitely, and it only takes a very small number of people to breed out with people from beyond their direct gene pool for that DNA to rapidly descend through the generations.

Chang factored that into a further study of common ancestry beyond Europe, and concluded in 2003 that the most recent common ancestor of everyone alive today on Earth lived only around 3,400 years ago.

He used two calculations, one that simply crunched the math of ancestry, and another that incorporated a simplified model of towns and migration and ports and people. In the computer model, a port has a higher rate of immigration, and growth rates are higher. With all these and other factors input, the computer calculates when lines of ancestry cross, and the number comes out at around 1400 BCE. It places that person somewhere in Asia, too, but that is more likely to do with the geographical center point from which the migrations are calculated. If this sounds too recent, or baffling because of remote populations in South America or the islands of the South Pacific, remember that no population is known to have remained isolated over a sustained period of time, even in those remote locations. The influx of the Spanish into South America meant their genes spread rapidly into decimated indigenous tribes, and eventually to the most remote peoples. The inhabitants of the minuscule Pingelap and Mokil atolls in the mid-Pacific have incorporated Europeans into their gene pools after they were discovered in the years of the nineteenth century. Even religiously isolated groups such as the Samaritans, who number fewer than 800 and are sequestered within Israel, have elected to outbreed in order to expand their limited gene pool.

When Chang factored in new, highly conservative variables, such as reducing the number of migrants across the Bering Straits to one person every ten generations, the age of the most recent common ancestor of everyone alive went up to 3,600 years ago.

This number may not feel right, and when I talk about it in lectures, it often results in a frown of disbelief. We're not very good at imagining generational time. We see families as discrete units in our lifetimes, which they are. But they're fluid and continuous over longer periods beyond our view, and our family trees sprawl in all directions. The concluding paragraph of Chang's otherwise tricky mathematical and highly technical study is neither of those things. It's beautiful writing, extremely unusual in an academic paper, and it deserves to be shared in full:

> Our findings suggest a remarkable proposition: no matter the languages we speak or the color of our skin, we share ancestors who planted rice on the banks of the Yangtze, who first domesticated horses on the steppes of the Ukraine, who hunted giant sloths in the forests of North and South America, and who laboured to build the Great Pyramid of Khufu.

You are of royal descent, because everyone is. You are of Viking descent, because everyone is. You are of Saracen, Roman, Goth, Hun, Jewish descent, because, well, you get the idea. All Europeans are descended from exactly the same people, and not that long ago. Everyone alive in the tenth century who left descendants is the ancestor of every living European today, including Charlemagne, and his children Drogo, Pippin, and, of course, not forgetting Hugh. If you're broadly eastern Asian, you're almost certain to have Genghis Kahn sitting atop your tree somewhere in the same manner, as is often claimed. If you're a human being on Earth, you almost certainly have Nefertiti, Confucius, or anyone we can actually name from ancient history in your tree, if they left children. The further back we go, the more the certainty of ancestry increases, though the knowledge of our ancestors decreases. It is simultaneously wonderful, trivial, meaningless, and fun.

This is not to disparage genetic genealogy and ancestry. Done right, DNA is an immensely powerful tool for studying families and the subtleties of human migrations over history. It can disclose unknown cousins or parents, and identifies the fathers and mothers of adopted children when records are lost or destroyed. Genealogy has been incalculably enriched with the fusion of genetics and the traditional forms of family detective work, such as using surnames or following the paper trails of births, deaths, marriages, and the documents that record the minutiae and particulars of a life. The problems are the same for both old and new techniques: The further back we go, the dimmer the past becomes.

The lure of fame or infamy in your family past is strong. That Christopher Lee is descended from Charlemagne adds to his charisma. Alas, it is also a hollow claim. He may well have carried none of Charlemagne's DNA, as indeed you might not, despite being equally related to that great emperor. On these vines of family history, there is thorny math that underlies all of these inevitable conclusions, and they are impenetrable to all but a handful of computational geneticists scattered round the world. We rely on the press and media to report and translate these complex stories, which are exclusively published in technical language in academic journals that are only intended to be read by other experts in the same fields. This is how scientific research is reported.

Often journalists do very well, and carve out narratives that the public can grasp and revel in without deviating from the implications of the actual data. Families and history mean so much to us, though, that we are frequently bewitched by the promise of historical eminence. Journalists can equally be lured into yarns for which the evidence is sketchy, simply because the narrative sounds exciting. When we look to the past and to our presumed genetic ancestry to understand and explain our own behaviors today, it is not much better than astrology. The genes of your forebears have very little influence over you. Unless you carry a particular disease that has passed down the family tree, the unending shuffling of genes, the dilution through generations, and the highly variably and immensely complex influence that genes have over your actual behavior mean that your ancestors have little sway over you at all.

Even so, with the advent of cheap genetic sequencing, a deep internal history of everyone can be revealed. We carry the traces of our ancestors in our cells, and nowadays, for the price of a second-hand copy of *Burke's Peerage*, the bible of genealogical record, you can pay to have your past supposedly unscrambled. Plenty of companies have emerged who provide this service, and I have had my genome analyzed by two of them, BritainsDNA (sic) and 23andMe. The kits are pretty similar: They come in a paperback-sized, high-quality parcel, and inside is a plastic test tube with a lid containing a sealed fluid. You are asked to fast for an hour beforehand, which allows your mouth to be mostly free of the DNA of the food you put in it, and then you are asked to conjure up quite a volume of saliva. This dribbles into the test tube, and you snap the lid shut, an action that punctures the seal and releases the preserving liquid into your spit. In all that frothy saliva are cells that have sloughed off the inside of your cheeks and gums, and within them is your genome. The mix preserves the DNA for the mail journey back to the lab. There, your DNA will be extracted, cleaned up, purified, and thousands of precise segments of it amplified and read. This isn't a whole genome analysis, like the Human Genome Project laid out in 2001, but a selection of some of the sequences that vary between people—the ones that have been already established to be of interest to science and to the paying customer.

Once the sampling is done, and a digital file created with your precise genetic read at all these interesting SNP sites in your genome, the data gets plugged into the database with everyone else's interesting sites, and you get compared to every other paying customer. Genetics is a probabilistic science, so the presence of a T rather than G at a specific location might indicate that you are more likely to have one condition over another. Some of these SNPs are informative for ancestry, some of them for physical characteristics such as hair color or eye color, or your predicted response to alcohol or caffeine, aversion to coriander, ability to smell asparagus in your urine, likelihood of baldness, tolerance to lactose, and, of course, the all-important wetness of your earwax. Some are associated with the risk of developing diseases such as Alzheimer's, types of aggressive breast cancer, or a particular type of emphysema.

These, it cannot be stated enough, are calculations of odds, not of destiny. Having Lionel Messi on your soccer team does not guarantee you victory in the World Cup finals, as Argentina discovered in 2014, despite him being the global player of the year. The genetic predictions are based on the frequency of these diseases in a whole population, and the very particular genetic sequence that occurs in those patients with those diseases. So, for example, the fact that my sequence came back without a SNP associated with Parkinson's disease does not mean I will not get Parkinson's disease. It means that my chance of getting Parkinson's disease with this particular gene variation is average. Conversely, according to 23andMe, I have a genotype that is of higher risk than most people for developing Alzheimer's disease. That does not mean that I will get Alzheimer's disease, it means that the chance I will is slightly higher than most people. Similarly, if you don't have that genotype, you are not immune to Alzheimer's. Knowing my own personal risk neither bothers me nor has prompted a change in my behavior at all.

For the physical characteristics, they are kind of interesting, in an instantly forgettable way. In my 23andMe readout, it confidently says that my eyes are "likely brown" due to the presence of an A in one version of a gene called HERC2, and a G in the other. An A and an A would make me more likely to have brown eyes, and two

Gs would make me lean heavily toward being a blue- or green-eyed man. I do have brown eyes, something that I had established a few years ago, with the less exciting technology of a mirror. Nevertheless, it is nice to have it confirmed at a molecular biological level. The fact that these characteristics are couched in these terms, though, shows a good understanding of the nondeterministic nature of genetics, and sarcasm aside, this is good science. These readouts are not necessarily predictive: They are effectively saying that most people with this particular genetic sequence have brown eyes, and we can therefore predict that you do too. It was interesting to see that 23andMe reported that I do not carry a version of the MC1R gene that is associated with red hair, but the company BritainsDNA did. This merely means that there are multiple alleles—alternative spellings—of this gene that bestow ginger hair on the owner (if you have two copies; I do not), and the two companies scan for different flavors.

When it comes to ancestry, the results demand even more scrutiny. The first thing worth mentioning is that I'm not entirely sure what we mean by *ancestry*. It's an imprecise word, and although an obvious and sensible working definition might be "the people from whom you are descended," we've already established that this in isolation is not very useful for the climb up your tree. It fully depends on what time period you are sampling. Many people think of ancestry in terms of ethnicity or geography. My father was born in Scarborough and my mother and her parents in Georgetown, Guyana. My ancestry one generation back is therefore northeast England and Guyana. My mother's grandparents were born in India, though. Each generation up, the number of ancestors increases, and the spread of where they were born does too. By the time it collapses back in on itself, say, 500 years ago, my ancestors were from all over Europe, and presumably all over the subcontinent and beyond. By 2,000 years ago, they are all over the world.

There are only really a couple of definitive unequivocal statements that we can make about your origins in the deep past.

The first is that 100,000 years ago we were all African. At that time, there were no *Homo sapiens* anywhere on Earth apart from in Africa, as far as we know. After the exodus from the motherland, however, it becomes more difficult to label people in these terms with such confidence.

The second is about Neanderthals. 23andMe allocated 2.7 percent of my genome to a Neanderthal origin, which is exactly average for most Europeans (though higher than the estimates from published academic science). I can say with some confidence that that DNA did not enter my genome more recently than 30,000 years ago, because we think that Neanderthals were extinct by then, and this introgression of their DNA into the genomes of us must have ended. As far as we know, Neanderthals never made it back into Africa (though some of their genes did, hitchhiking on Eurasians making a return journey). We can predict with some degree of accuracy when particular SNPs or collections of SNPs originated, but there are millions of these, and they emerge into our genomes at different times in history. Where geographically they emerge is virtually impossible to work out, because of the constant flux and movement of people around the world. The everyday, real world use of the word *ancestry*, not for the first time, doesn't quite align with the precision that science asks of it.

It's important to remember that the commercial DNA ancestry tests don't necessarily show your geographical origins in the past. They show with whom you have common ancestry today. I have a few chunks of DNA that 23andMe showed are most common in Scandinavia now. Vikings were from Scandinavia a thousand years ago. Am I descended from Vikings? Yes! But not because of this analysis. It's because everyone in Europe has Viking ancestry. Admittedly, the proportions do vary between people and these differences correlate with geography. It's within these subtleties that genetic ancestry is a robust and useful science. We can see the slow migrations of people, and we can see the ghosts of invasions and similarly their absences, as shown in the formidable People of the British Isles study (page 97). In general, though, the mix over 1,000 or 10,000 years will reveal very little about you.

BritainsDNA is a company that generates a lot of media coverage with the claims that come alongside the selling of their genetic ancestry kits. In the last few years, there have been plenty of high-profile stories in the press in the UK that purport to reveal breathtaking and breathtakingly specific stories as revealed by DNA analysis. They were behind the dodgy claim in 2014 that people with red hair are all set for extinction due to the effects of climate change (as debunked on page 90). They were the testing company who provided the data for the front page splash in the *Times* that heir apparent Prince William harbored Indian DNA, and this will one day be the first of its type to be found in the top tier of English monarchy.*

In 2012, BBC Radio 4's main morning news program *Today* featured an interview with the chief executive of BritainsDNA, Alistair Moffat, in which he asserted that their tests had revealed descendants of the Queen of Sheba. The Queen of Sheba, by the way, is a character in a story from the Bible, and there is no physical

* This claim is built on unsteady foundations in itself. The superlative genetic genealogist Debbie Kennett thoroughly deconstructed the assertion, which is based on the type of mitochondrial DNA that Prince William bears, as sampled in one of his cousins from the same maternal lineage. This DNA is a fairly rare type, and may or may not be associated with India. But because the data was not published in a scientific journal, instead simply injected into a newspaper without the scrutiny of peer review, it's impossible for us to know. Debbie Kennett, who is thorough and knowledgeable, and very well respected in the genealogy community, concluded that the *Times* "has compromised its integrity by publishing an advertorial."

evidence of her actual existence. She features in many diverse legends from the Middle East over many centuries. In 1 Kings 10 in the Old Testament, she rocks up at the palace of King Solomon with a train of camels laden with gold and spices and other trinkets, but we're not sure where she is supposed to have come from; Sheba might be Saba, a place thought to have existed in what is now Yemen. Can we say she definitely existed? Not really. If she did, can we say that anyone is descended from her? I wish I could give a more academic answer, but a frown and a shrug is the best I can manage.

Celebrities are regularly pulled into this web too. The British comedian Eddie Izzard featured in a jovial documentary series on BBC1 in 2013 in which he trudged all over the world using snippets of his genome as a guide as his ancestry was supposedly traced. We met many interesting people on this trip, which charted humankind's journey out of Africa all the way to Eddie, using some genetic half-truths as the map. The final conclusion, filmed in Denmark, was that he had Viking ancestry, due to the presence of a particular sequence of DNA in his mitochondrial genome, which is inherited solely down the maternal lineage. Among a battery of questionable assertions, we were told on the program by Jim Wilson—academic geneticist and cofounder of BritainsDNA—that this particular type of DNA is 2,000 years old, which is a bit puzzling given that the Vikings didn't exist 2,000 years ago. It can't be stated enough that there is no test for Viking DNA, but I can assure you that if you are white, you have Viking ancestry.

In April 2012, the actor Tom Conti was in the news as being of prestigious genetic ancestry. The *Daily Telegraph* reported that Conti's

> ancestors settled in Italy around the 10th century before one of them, Giovanni Buonaparte, settled in Corsica and

> founded the family branch that produced Napoleon . . . He [Conti] is clearly a close relative of Napoleon. Only DNA could have told that story.

The DNA in question is a type on the Y chromosome, so only passed down the male line, called E-M34. It's common in Ethiopia, the Near East, and Europe. A French study in 2011 plucked three beard hairs from a verified collection of Napoleon's possessions, and extracted DNA from the cells clinging to the base of these whiskers. Indeed, the Y chromosome type is E-M34. They're related in the same sense as millions of other men with this type of Y sequence are, and it originated several thousand years ago. Furthermore, as with Charlemagne's descendants, it's compelling to see named individuals on documented family trees that date all the way back to the middle ages. But we know, via math and genetics, that your ancestors were also settled in Italy in the tenth century, regardless of whether you're Tom Conti, Eddie Izzard, President Obama, Richard Dawkins, Taylor Swift, Adolf Hitler, Pope Francis, Queen Elizabeth II, Madonna, Maradonna, Rabbi Jonathan Sachs, all four members of ABBA, my butcher, or Charles Darwin.

And so it goes on. BritainsDNA (which also trades as ScotlandsDNA and CymruDNA in those other British domains) has a long record of attracting gullible media coverage with historical claims, many of which are somewhere between speculative and unsupportable. The only way we could ever definitely say where your genetic origins are a thousand years ago would be to dig up the bodies of everyone who lived a thousand years ago, and then compare them. And the answer would be that your genetic origins are, in fact, in everyone.

When my results came back from BritainsDNA, there was a nice line drawing of my ancestry "type" as defined by my particular

Y chromosome sequence: "Germanic"! An angry topless fellow with a beard and a fulsome ponytail and drawstring trousers, bowling forward with a shield and short-sword, looking very cross indeed.

> Your S21 YDNA marker is GERMANIC, and you are descended from the peoples of the Rhineland and the Low Countries. They first reached Britain by a remarkable and little understood route only available to prehistoric peoples. And much more recently, many people with your marker arrived in the historic period.
>
> As the Roman Empire in the west convulsed and collapsed in the 5th century, many Germanic peoples migrated, some crossing the Alps to Italy, others sailing the North Sea to Britain. In the very cold winter of 405/406AD the Rhine froze and the Vandals, Swabians and Alans skidded across the ice, broke through Rome's flimsy frontier defences and ravaged the Roman province of Gaul. Perhaps as many as 70,000 warriors walked across the river. Your genetic cousins were part of this tumultuous period of Europe's history, what German historians call Die Volkerwanderung, the Wandering of the Peoples.

Skidding Alans!

Goodness, that is very exciting, although it is disappointing that my ancestors committed the fashion travesty of wearing drawstring pants. "Germanic" is BritainsDNA's own category, not one used in history or science. It's pretty vivid, and quite appealing if you don't apply much scrutiny. But it's also meaningless. The Y chromosome makes up less than 2 percent of my total DNA. Does my haplotype mean that I am descended from a group of people from northern Europe in the last two millennia? Quite possibly. The map of the distribution today of this particular type of DNA in my BritainsDNA results shows it most densely crammed into the people of the Netherlands, they claim before 1500 CE (though I am unaware of how they could know this: It is possible, but being a commercial company, their data is not publically available, nor are their methods). That type of Y chromosome is also present today from Svalbard to Gibraltar to Vladivostok.

The Y is a tiny proportion of the total DNA I possess, and in fact less than the amount of DNA that I and most Europeans have inherited from Neanderthals, at least according to the rival DNA testing company 23andMe. To label my "ancestral type" as this Germanic warrior with all his gliding across the frozen Rhine in unfashionable trousers is absurd. By simple percentages in my genome, I am more Neanderthal than this bearded character.

Another tiny bit is from my mother's lineage, the mitochondrial genome, which was not on the database of BritainsDNA at the time of my test, as these types of company generally add data as they add customers. 23andMe report that it is most common in India—again, not a tremendous surprise given that my mother is Indian. The mitochondrial genome harbors just 37 genes, and the Y chromosome 458. There are around 20,000 genes in humans, and other than those 495 that are exclusive to paternal and maternal lineages, the rest are housed on the autosomes—the 22 pairs of chromosomes inherited from both parents. The source of those is everybody you are descended from, mixed and remixed and shuffled and reshuffled every time a sperm or egg was made in one of your millions of ancestors.

Some companies offer services whereby they will tell you the supposed precise village location of your genetic ancestry a thousand years ago. Again, it's a peculiar thing to claim, as you will have millions of ancestors a thousand years ago, and it doesn't take an Early Middle Ages historian to be confident that they won't have all come from the same village. As the work of Chang, Ralph, and Coop, and many others has shown, the tendrils of ancestry that sprout upward from you become unfathomably enmeshed the further back you go, until all reach all people a thousand years ago. The genealogy is inferred, and much of the DNA from all these ancestors is not present, so a single geographical location is little more than meaningless. The calculation averages a location based on where you share some bits of DNA in the present day, and its early versions clearly needed some work: It placed the "genetic origin" of one paying genealogical hunter customer, one Julie Matthews from Maryland in the US, in the tidal waters of the Humber estuary.

Our desire for these stories fuels the companies that sell them. Your people were Germanic wandering warriors! You're descended from the Queen of Sheba! Your DNA was forged in this Middle Ages village! You're related to Napoleon!

They just tell a tale: largely unsupportable, possibly true, or generically accurate for millions, and underwritten by thinly stretched DNA data. The Forer effect is a psychological phenomenon where people conclude that broadly true statements are accurate for themselves personally, when they are in fact generically true for many people. In 1948, psychologist Bertram Forer gave his students a personality test, and followed each one with a bespoke vignette of their character. They were then asked to rate their personalized analysis, which they did very positively—an average of 4.26 out of 5. Except they weren't personalized at all. All the personality sketches were identical, made up of thirteen bland statements that vaguely describe common or desirable personality traits:

> At times you are extroverted, affable, sociable, while at other times you are introverted, wary, reserved.

This is how astrology works—divination drawn from banality, that we are complicit in. We cling to the things that appeal, and happily ignore the rest. This is the basest sentimentality. How odd that our ancestors are never identified as dimwits, or shoe salesmen, or turnip peelers. They're always fearsome warriors, or deer hunters, or Saracens. In the gold-plated phrase of Mark Thomas from UCL, much of the business of ancestry that utilizes our DNA is "genetic astrology." Everybody loves a good yarn from history, and we are all hungry for tales of our own ancestors. That's all well and good, as long as we recognize it for what it is.

What does this all mean? The truth is that we all are a bit of everything, and we come from all over. Even if you live in the most remote parts of the Hebrides, or the edge of the Greek Aegean, we share an ancestor only a few hundred years ago. A thousand years ago, we Europeans share all of our ancestry. Triple that time and we share all our ancestry with everyone on Earth. We are all cousins, of some degree. I find this pleasing, a warm light for all mankind to share. Our DNA threads through all of us.

Ancestry is messy and difficult. Genetics is messy and mathematical, but powerful if deployed in the right way. People are horny. Lives are complex. A secret history is truly hidden in the mosaics of our genomes, but *caveat emptor*. No scientific test exists that will tell you where the DNA that you would come to inherit was precisely located in the past. Human history is replete with the fluid movement of people, and tribes and countries and cultures and empires are never, ever permanent. Over a long enough timescale, not one of these descriptions of historical people holds steadfast, and only a thousand years ago your DNA began being threaded from millions from every culture, tribe, and country. If you want to spend your cash on someone in a white coat telling you that you're from a tribe of wandering Germanic topless warriors, or descended from Vikings, Saracens, Saxons, or Drogo of Metz, or even the Great Emperor Charlemagne, help yourself. I, or hundreds of geneticists around the world, will shrug and do it for free: You are. And you don't even need to spit in a tube—your majesty.

ii: Richard III, Act V

SCENE III. Bosworth Field.

. . .

RICHARD III: . . . every tale condemns me for a villain . . .

SCENE V. Another part of the field. Alarum. Enter KING RICHARD III and RICHMOND; they fight. KING RICHARD III is slain. Retreat and flourish. Re-enter RICHMOND, DERBY bearing the crown, with divers other Lords

RICHMOND: God and your arms be praised, victorious friends,
The day is ours, the bloody dog is dead.

Shakespeare's story doesn't quite end there. Richmond lauds his final, inevitable victory at Bosworth Field, marries Elizabeth, and the Wars of the Roses are over. "England hath long been mad, and scarred herself," says Richmond, soon to be crowned as Henry VII, and the "the white rose and the red" are united, the schism between the Houses of Lancaster and York healed in genetic recombination through breeding. So ends Act V of *Richard III*, and his reign of murderous terror.

Exeunt.

Just like the killer in pretty much every horror movie, he was due one more dramatic entrance, the epilogue hand thrust out of the grave before the credits roll. We would have to wait another five centuries to watch Act V of *Richard III* play out.

This is Shakespeare's Richard, of course, an utter bastard of a king. He's culturally ubiquitous as the baddest of Shakespearean villains, twisted morally and spinally, his death predicted throughout the play, and he's dispatched by the ascendant Richmond unceremoniously, and notably without the option of a grandstanding death speech. We often think of him as in an iconic depiction, the Society of Antiquaries of London Arched-Frame Portrait. In it he is pale-faced with bobbed hair, a stern look, fiddling with his ring. Was it his likeness? That painting is from the sixteenth century, and the earliest known portrait follows his death by at least a quarter of a century, so it's difficult to know. More often, we think of him portrayed by Laurence Olivier, thin lipped, humped and lame-armed, bombastic and sneering:

> Now is the winter of our discontent . . .

What do we know of the real son of York? Richard was born in October 1452, the twelfth child of Cecily Neville and Richard Plantagenet, the Duke and Duchess of York. The climate was one of political turmoil that would rage during his life, and bring about the reign of the Tudors at his death. Edward IV had had two goes at kinging, the first from 1461 to 1470, and then again from 1471 to 1483 after a slight return of an enfeebled Henry VI. Edward IV dropped dead on April 9, leaving the crown to his twelve-year-old son Edward V, but it never made it onto his head. Richard, brother to Edward IV, uncle to the V, was appointed his Protector while he was underage.

But Richard had other plans. He escorted his nephew to the Tower of London, where he was soon joined by his other nephew, Richard of Shrewsbury, under the promise of his brother's coronation. While they were there, Richard declared the marriage of their parents to be invalid on contractual grounds (which sometimes happened in those days when royal conflicts raged). On June 22, 1483, the two princes were declared bastards, and Edward V was thus ineligible to rule anything. Richard was crowned King of England fourteen days later. The two boys, forever romanticized as the Princes in the Tower, were not seen again after that summer, nor their remains ever unearthed. Whether Richard III had them executed or not is unknown, and historians debate it to this day.

His reign was neither long nor stable. A York rebellion followed in 1483, and Henry Tudor, then in exile, was proposed as the rightful king. Skirmishes and battles characterized the two years before they met on Bosworth Field, where Henry's men, though outnumbered, circled Richard, and took him out.

> And thus I clothe my naked villainy
> With odd old ends stol'n out of holy writ;
> And seem a saint, when most I play the devil.

It was August 2, 1485, and Richard was thirty-two. His body was stripped naked and paraded around, carcass slumped over the rump of a horse for all to see and jeer, a king humiliated and trounced. The manner of his death was reported to be brutal: According to the French chronicler Jean Molinet, while Richard's horse was quagmired, he received a strike to the back of his head—un-helmeted, possibly to rally troops, to show he was leading from the front—from a halberd, a long two-handed axe-headed spear. Perhaps this equine immobility gave rise to the Shakespearean legend of his pleading a trade "my kingdom for a horse" to escape the thick of battle. That halberd blow, or other head wounds, were death strikes, and Richard III instantly became both the last of 300 years of Plantagenet rulers, and the last English king to die on the battlefield. England was different after that; Henry VII reigned till

1509, who gave way to the VIII, and then the latter's children, Edward, then Mary, and then the first Elizabeth.

After being paraded around, Richard's corpse was taken to the churchyard of the Greyfriars in Leicester and inhumed. Details are scant, but it was thought to be a hasty burial with no fanfare. A decade later, Henry VII paid for a headstone of sorts, a £50 alabaster marble for his fallen predecessor. However, the next Henry revolted against the Vatican and severed ties with the Holy See. From 1536 onward, he dissolved the monasteries, a process that reached Leicester two years later. And that was the end of Greyfriars. Richard's penultimate resting place was erased from memory or record, the monastery bought, demolished, and the masonry sold.

We've been looking for Richard in some form ever since. *Richard Tertius* by Thomas Legge is quite probably the first English historical play in 1580, though Tudor propaganda by Thomas More and others, and especially Holinshed's *Chronicles*, had established myths about him long before. Ben Jonson wrote *Richard Crookback* in 1602, though it was never published. Of course, Shakespeare's 1594 version endures above all. Acting titans have sought to play him in all his twisted menace, from Alec Guinness to Kenneth Branagh, and even John Wilkes Booth, the actor who later assassinated Abraham Lincoln, who was himself a known fan of both Shakespeare and the play; and on screen, Olivier in 1955, Ian McKellen and Al Pacino, and most recently Benedict Cumberbatch on the BBC.

But Richard's greatest entrance came in September 2012.* The site of the dissolved monastery was well known to local historians, and the area is still known as Greyfriars. Leicester is replete with

* Many people have driven this exciting endeavor forward, and I haven't specified who did what above in this truly impressive project. Philippa Langley from the Richard III Society instigated the search, Richard Buckley headed the dig and coordinated the project overall, Turi King led on the genetics, Jo Appleby the osteoarchaeology, John Ashdown-Hill the identification of Richard's descendants. Here is a full list of the authors on the scientific publication

Ricardians. Since 1924, the Richard III Society has sought to reassess his reign, which is characterized mostly by Shakespeare's dastard. In August 2012, a team from the University of Leicester, with Ricardians and geneticists in tow, announced that they were on the hunt for the grave. Paper trails in the couple of years before his comeback were inconclusive, and similarly radars that bounce back from underground structures only revealed pipes and other public utilities. But of the various methods of hunting for a medieval grave, the best guess was that beneath a Social Services car park lay late medieval structures of the friary, and that these would be a good place to dig for the king. So on August 25, they dug.

Success was bafflingly immediate. Anyone who has ever been on an archaeological dig will know that it is beset by days and weeks of sifting through mud and dust, spooning crumbs of dirt that hide the decomposing clues to the long dead or to past lives, which could be easily ignored or obliterated forever with a lusty spade. On day 1, they found a pair of skeletal legs, missing feet.

The legs were covered while trenches were dug over the next few days, which uncovered structures of the monastery. As is required in the UK, the Ministry of Justice gave permission to exhume, and the skeleton was fully unearthed, and removed on September 6. There was no trace of a casket or even a death shroud, and the head was cocked uncomfortably upward, indicating that the dug hole was too small, that is, inconsistent with a formal ceremonial burial. Whoever this was, he was slung in a hole in the ground. The hands were placed together on the right hip, which might indicate that they were tied, though no traces of binding were detected. The

---

entitled "Identification of the remains of King Richard III" (2014): Turi King, Gloria Gonzalez Fortes, Patricia Balaresque, Mark Thomas, David Balding, Pierpaolo Maisano Delser, Rita Neumann, Walther Parson, Michael Knapp, Susan Walsh, Laure Tonasso, John Holt, Manfred Kayser, Jo Appleby, Peter Forster, David Ekserdjian, Michael Hofreiter, and Kevin Schürer.

spine was notably curved, not as a result of the haphazard posture of burial, but pathologically a crookback.

> Sin, death, and hell have set their marks on him,
> And all their ministers attend on him.

Scientists, not ministers, attended on his death the second time round. And so began the next phase of identification. The bone morphology suggested a man, probably five feet eight inches based on the length of the femur (the thighbone). The sex is something that DNA would confirm beyond doubt. As ancient DNA does not preserve the neat chromosomes of sex in a visible way, we can fish around in DNA and tease out fragments that are exclusively found on those two chromosomes. The presence of DNA found only on the Y confirmed the bones were that of a man.

The curved spine shows classic scoliosis. Normal backbones are gently S-shaped, but only if you look from the side. Viewed from the front or back, they should be plumb vertical. Scoliosis is a condition, either inherited or acquired during life, when the spine curves to the left or right, like a C or in extreme cases an S. That can have the effect of making the owner of the bowed spine carry one shoulder higher than the other—not exactly a hunchback, but certainly a crook.

The skull, well, that was a ruin. In the grave it was positioned uncomfortably but whole, at least from the front at first glance. The discomfort of the burial was of no consequence though; viewed from the back, this was a man whose head wounds were brutal and fatal. Feel your own head with your right hand at the point just above where your neck meets the skull. This section of skull is called the occipital, and covers the protuberance at the back of the head, as if you were cupping a hand across a bony outcrop. The Greyfriars skull had a hole sliced away on the lower right-hand side of the occipital wide enough to comfortably fit a plum through it.

An injury like this could be caused by the blade of a halberd, or a sword, and would have exposed and diced a decent chunk of the brain, and that would be instantaneously fatal in itself. Imagine the force of a blow that would deliver this wound.

On the other side in roughly the mirror spot is a jagged hole caused by a sharp weapon, possibly a sword or the pointy end of a halberd. It had been thrust upward with such force that it penetrated through his entire brain, and scored a mark into the inside of the skull case. That too, would have caused instant death, and indeed neither of these holes shows signs of the process of wound repair. It's not possible to say whether they occurred just before death or just after; *perimortem* is the word we use—around death. But either would be mortal, and neither is consistent with wearing a helmet. This man died with his hat off.

Leicester has a deep-rooted history of genetics. In 1984, Sir Alec Jeffreys invented the techniques known as DNA fingerprinting that have been used to finger criminals ever since. For Richard, the genetics fell to a scientist with a suitably regal name, Turi King. In cases of the attempted identification of a named person, you are asking the question "is this person who we think it is?" Noble family trees are the best documented of all, and so it is right and proper to rely on the two genetically thin pillars of ancestry—the Y chromosome and the mitochondrial. Comparing Y DNA to family trees is frequently a less arduous piece of detective work as it's passed only through men, as for the most part are surnames.

The family tree of the houses of York and Lancaster are well drawn, and the Richard Project managed to find men descended from Richard down the male line. It starts with Edward III, Richard's great-great-great-grandfather, who ruled England for fifty years in the fourteenth century, including during the decimation by the Black Death. His Y chromosome was passed down to Richard, where it stayed, but also wends its way via Henry the 3rd Duke of Somerset, to Henry Somerset, the 5th Duke of Beaufort in the eighteenth century. From there, they found five living men who, though they are not willing to be identified themselves, were willing to spit in a tube to parse their Ys. The trouble was, if the Greyfriars skeleton was Richard III, the genetics and the family trees were not in agreement. The five descendants did not have the same version of the Y as the skeleton. Men are notoriously unreliable in matters of sex, as our contribution to conception is over in a flash, and therefore the possibility of false paternity is very real. Estimates vary of how many presumed fathers were not present in that flash, but in the case of the identification of Richard, Turi King used a figure of around 2 percent—out of a hundred people, two of them were not sired by their father. There are nineteen generations from Edward III to Henry Somerset, which puts the possibility of false paternity at a chance of about one in six.*

Case closed? Not so. Women to the rescue. False maternity, for fairly obvious reasons, is a lot less common. Women are more easily lost to history, not least as their surnames traditionally have been abandoned with marriage. They pass on their mtDNA in an indelible way though. So Richard's elder sister, Anne of York, carried the same mtDNA as him, passed down from their mother, Cecily Neville. Around twenty generations from her brings that

* The sharp-eyed will note that this Y chromosome mismatch implies that the false paternity might illegitimize the Tudors' claim to the throne. Where the break occurred is unknown, and could only be found by digging up a lot of bodies from John of Gaunt (1340–99, the last common ancestor of both Richard III and Henry VII) to the present. Nevertheless, technically, as the Windsors are directly descended from the Tudors, the British Monarchy may yet turn out to be built on a falsehood, albeit a very old one. One of the authors, historian Kevin Schürer, pointed out to the *Guardian* at the time that the "Tudors took the crown because they killed Richard at the Battle of Bosworth in 1485, not because they could prove the blood royal flowed through their veins." Buckingham Palace is yet to comment.

very same mitochondria into the present. Two maternal descendants were found via family trees: a Canadian named Michael Ibsen, and one Wendy Duldig. The mitochondria is a small loop of DNA only 16,500 bases long. That's still plenty enough for variation that can split the difference between two different matrilineages, and so Turi King and her team performed a battery of comparisons between the mtDNA of Ibsen and Duldig, and the bones in the car park. They first looked at regions in the DNA that are known to be hypervariable* and, just to be sure, extracted the ancient DNA twice and compared it in two different independent labs. The regions were identical. They sequenced the entire mtDNA genomes, and found that Ibsen's was identical to the bones, and Duldig's differed by just one letter. That's still perfectly consistent with them being from the same lineage. Just to be sure, King calculated the odds of these being coincidences, by comparing other bits of the mtDNA of the three of them and a database of 26,000 other mitochondria, with the result that there were no crosses. And, to be on the safe side, all members of the team had their mtDNA checked, and all the men their Ys, in case they were a contaminating source. No dice.

And so it was him. The men in his family tree were not much use at giving up genetic evidence, but the women were rock solid. The rest of his genome, aside from the mtDNA and the Y, provided just a touch more detail. Richard bore variants of two genes that are most frequently seen in blue-eyed blonds. But it's a genotype that varies with age, and blond babies often become darker haired adults. The "Arched-Frame Portrait" is the depiction that most aligns with how we try to predict what hair and eye color results from Richard's kind of DNA.

* These are known as hypervariable regions.

It's important to stress that DNA was a linchpin in the identification, but only part of the evidence. The bones match the best contemporary descriptions. The scoliosis, the head wounds sound like what the historians say, though not what Shakespeare describes; there is no withered arm swinging limply from the crooked shoulder. But the shoulder was crooked. Other pathologies, such as subtle gouges to the hips and waist are not consistent with an armor-clad victim, which chimes with the reported humiliating naked parade. And the skull is a fierce reminder of the reality of war, one or two formidable blows that would have destroyed brain activity in a heartbeat. All of that says Richard. But the DNA in consort with the paper trail of a genealogy available only for royalty says this was him. Richard III is now the oldest person to be unequivocally identified in death.

The misidentification of Jack the Ripper

That's how you do it. These types of study are akin to a missing person, and with the case of Richard III the bar has been set appropriately high. Compare this great regal success to another nefarious celebrity detective story that filled the world's newsfeeds and pages briefly in 2014. On September 7, we awoke to discover that the mystery of Jack the Ripper had finally been solved. In 1888, in the Whitechapel district of London, the bodies of five women were discovered between August 31 and November 9. Four of them had been anatomized gruesomely and apparently ritualistically. This befuddled an incompetent and baffled police force at the time, who were also taunted by the murderer, or by hoaxers, with graffiti at the crime scenes, and a letter, addressed "From hell," which came with half a human kidney; the other half, the letter claimed, had been eaten.

These hideous crimes have spawned a whole morbid industry that somehow sidesteps the fact of the profound and disturbingly grotesque violence he perpetrated against five vulnerable women. Ripperology is a thing now, where amateur sleuths propose theories on the identity of the Ripper, and it's big business: Amazon lists more than 4,000 books if you search with the words *Jack the Ripper*. There are international conferences, and in 2015 the Jack the Ripper Museum opened in Whitechapel, to widespread criticism, protest and some vandalism. Even the museum's designer decried it as "salacious, misogynist rubbish," and in the gift shop you can buy all sorts of gaudy gimcrack, including for £45 a Ripper top hat.

He wasn't the first serial killer, but is one that coincided with the birth of mass media, and set the ghoulish tone for our modern obsession with murderers. His identity remains unknown. Dozens of men have been suggested, with dozens of wackaloon theories to sit alongside them. Suspects range from the ridiculous, such as Lewis Carroll and members of the royal family, to a more credible shortlist of six men. Amateur detectives pore over the forensic reports, and any snippet of data that might reveal a clue. Some fly off into the land of wild speculation, or join dots hopefully, or ponder the lives of the victims and protagonists under the misapprehension that by imagining their characters, their behavior will unveil the key that unlocks the mystery.

It's a strange, troubling business that we have become so obsessed with this serial killer, and romanticized his murders to a Victorian horror story, with its foggy east London scenery, and fallen women preyed upon by a cloaked, top-hatted villain. Whoever he was, he murdered women, which doesn't seem to me to be something worth celebrating. Of this obsession, none of the theories has provided any sort of evidence that could constitute a positive identification that would satisfy a historian, policeman, forensic scientist, or jury. Would DNA be the missing piece in this wicked puzzle?

In September 2014, that was precisely the claim. The *Mail on Sunday* prompted a global whirlwind of coverage with an extract from a new book by a self-professed "armchair detective" named Russell Edwards. The book has a hefty and explicit title *Naming Jack the Ripper: New Crime Scene Evidence, A Stunning Forensic Breakthrough, The Killer Revealed.* Inspired by watching the wretched Johnny Depp film *From Hell,** Edwards acquired a shawl at an auction in 2007, found, he claims, at the murder scene of the fourth woman, Catherine Eddowes. Edwards then approached Jari Louhelainen, a forensic geneticist at Liverpool John Moores University, to see what he could find. As a result of the DNA evidence supplied by Louhelainen, Edwards asserts that science has solved it "beyond reasonable doubt." The Ripper was indeed one of the known suspects—one Aaron Kosminski. He was a Russian Polish Jew who came to London in 1881 and worked as a barber in Whitechapel. In 1891, after the murders, his behavior had become so problematic that he was interned in a lunatic asylum. He didn't wash and picked up food waste from the street and ate it, and his medical notes suggest compulsive masturbation as the cause. He spent the rest of his life in asylums and died emaciated from gangrene aged fifty-three in 1919.

The forensic report of how Catherine Eddowes had been dissected is truly sickening, and there is nothing to be gained from indulging such a description of violence against women in order to entertain. Eddowes had been a casual sex worker, though she had other jobs, such as hop picking in the summer before her death. She was fairly destitute by the night of her murder, having split the

* Which bears little resemblance to the book from which it is nominally derived. *From Hell* by Alan Moore and Eddie Campbell is, as far as I'm concerned, a masterpiece, and the only work worth reading on Jack the Ripper. It is fiction, and a graphic novel, and meticulously researched essential reading on the murders and the media.

last of her money with her partner and pawned his boots. On September 29, she was taken into police custody after being found flat-out drunk in the street, but released at one o'clock in the morning of the 30th. Three quarters of an hour later, her mutilated body was found on Mitre Square in the City of London. The police surgeon, Dr. Frederick Brown, arrived on the scene around 2 AM, and wrote a detailed description of the scene, including the extent of her injuries. There's no explicit description of her clothes, other than to mention how they had been arranged in relation to the many incisions on her body.

I am not concerned with who Jack the Ripper was. I am concerned with how we can know. There is no kind way of saying this: Russell Edwards' front-page splash identification is a forensic fantasy from start to finish.

Aaron Kosminski was proposed as a suspect in the Jack the Ripper case in an 1894 police memorandum, but there's plenty of confusion about him and other names associated with the case and with the asylum. A "David Cohen" is mentioned in some records, which may have been a lazy pseudonym given by officials who struggled with a foreign-sounding Jewish-Polish name. The same goes for Nathan Kaminsky, who other authors have suggested is an alter ego, or a suspect himself. It doesn't really matter for our story, because the claim by Russell Edwards was that DNA had proved it was indeed Kosminski.

The basis for the claim is this: Edwards took the shawl to Louhelainen, who managed to extract DNA from it, and sequence parts of the mitochondrial genome. Using genealogy, they had managed to identify descendants of both Eddowes and Kosminski, and they matched. Kosminski's DNA was on Catherine Eddowes shawl that she had with her at the time of her death. As Edwards wrote in the *Mail on Sunday*, "DNA evidence has now shown beyond reasonable doubt" that it was Kosminski. Case closed.

Not even close. The first problem is that there is no report of the presence of the shawl at the scene of the crime. It was acquired by Edwards at an auction in 2007, and the only verification of its provenance was a letter from the man who sold it, who claims that it was acquired by an acting sergeant in the local police. He is not known to have been at the crime scene, nor was it on his beat. Instead he had supposedly asked his superiors if he could take the shawl home to his wife, a dress maker, and it had been passed through the family ever since. Edwards himself notes the unlikeliness of this fable: "Incredibly, it was stowed without ever being washed." He goes on to state explicitly in the splash "world exclusive" feature that there "was no evidence for its provenance."

Next problem is the question of why Kosminski's DNA was on the shawl. Edwards states that they had found traces of semen, suggesting that he might have ejaculated on it at the scene. We also have no evidence that the Ripper did that, and many serial killers do not, though their crimes look sexually motivated. If it were true, given Eddowes' work as a prostitute, might it have been that he was a client of hers?

But none of that matters anyway, because the DNA analysis itself was incredible—in the sense of being not credible. They didn't pull off any sperm from the shawl, but did find evidence of epithelial cells, the ones that line any wet surface in a human, and that therefore might travel alongside the man's vital fluids. But no autosomal DNA was found,* only mitochondrial. Contamination is a constant threat in any case that involves DNA, whether it's fresh today or from the prehistoric. There are carefully observed protocols for avoiding and eliminating the acquisition or identification of DNA from modern bodies. There's a picture in the *Mail on Sunday* of Russell Edwards holding the shawl up from two corners with his bare hands. This is not a good start. Louhelainen did tell me that they had checked against his and Edwards' DNA, but the descendants of Eddowes have been reported to have been in the same room as the shawl on one occasion, which is a potential source of their DNA sticking to the material. We shed DNA all the time. Your DNA is on the pages of this book, mine on the keys on this keyboard and in the microphone of my mobile phone, as it's even transmitted in minute, but detectable quantities, on our breath.

That isn't even the main problem. Louhelainen and Edwards' case rests on the mtDNA they could identify, as did Richard III's. The Ricardians sequenced the entire mitochondrial genome, whereas the Ripperologists looked at one particular variation, a mutation known as 314.1C, found in the DNA pulled from the shawl, and compared it to a standard DNA database. Louhelainen told the *Independent* newspaper that "it is not very common in worldwide population, as it has frequency estimate of 0.000003506, i.e. approximately 1/290,000." This formed the basis for the claim that it was unlikely that it could have come from any source other than Kosminski. Even so, that rarity might not be enough. I interviewed Louhelainen at the time, and asked him if he would be happy to present the evidence in a court of law if the crime had been committed in the last month: "You probably know the answer," was his reply, "this is based on mitochondrial evidence—it wouldn't be conclusive in the modern court."

The odds were wrong too. Sir Alec Jeffreys, the father of DNA fingerprinting, pointed out that the database of mtDNA they were comparing it against only has 34,617 entries, therefore "it is impossible to arrive at an estimate as low as 1/290,000." Maybe he put the decimal place too far right, and in fact he meant 29,000.

But that isn't even the problem either. In fact, the mtDNA mutation they were using was mislabeled. Instead of 314.1C, it was 315.1C, which more than 90 percent of people of European descent have. Any one of most people who might have had contact with this piece of material could have left it there.

So the lesson is clear and simple: Don't believe the hype. DNA is a clue, not a silver bullet. It takes skill and care to make that clue into something valuable. The cases of Richard III and Jack the Ripper are at polar opposite ends of the scientific spectrum in terms of skill and care. The royal project was meticulous, thorough, and careful, and used multiple strands of evidence, including radio carbon dating and forensic anthropology, and tied it all together with some precise DNA ancestry. Doubt is the cornerstone of science. It gives you a sure footing from which to build an argument based on data. This was bound by Turi King's statistical techniques that placed the level of doubt in the identification of his remains at thousands to one against. It was conducted in the glare of publicity, with a Channel 4 television crew omnipresent to record every aspect of the unfolding story. This type of company can pervert the scientific process, and certainly vexes some scientists. But even in this understandable media glare, the diggers expressed appropriate caution. The body found on the first days was indeed the king, but this was not confirmed for three months. The results went through the most rigorous and expected channels of peer review to enter the scientific literature in a well-regarded journal, *Nature Communications*, many months after the initial fanfare of exhumation.

None of these things were observed in scrutinizing the Ripper. Edwards told the *Mail on Sunday*:

> I've got the only piece of forensic evidence in the whole history of the case. I've spent 14 years working, and we have finally solved the mystery of who Jack the Ripper was. Only non-believers that want to perpetuate the myth will doubt. This is it now—we have unmasked him.

Then call me a nonbeliever. The provenance of the shawl is extremely sketchy. The handling of the shawl was a comical distance from providing reliable genetic data and precluding contamination. And the data analysis? Well, I've no idea, because it's never been published, but we do know it was faulty. Was Kosminski Jack the Ripper? Who knows? This is a missing person case that remains unsolved. The gloss of science won't help if the rest of the evidence is wafer thin. And if you don't publish in a way that others can check, you've simply got no game. This is not how we do it.

The inhumation of Richard III, Part II

The first time around, Richard was buried with "no funeral solemnity," according to the sixteenth-century scholar Polydore Virgil. The second time it was pomp all the way. Proceedings went on for the days between March 22 and 27, 2015, and a curious public queued for hours to gaze upon his coffin—built by his descendant Michael Ibsen, and encased in a tasteful Swaledale stone tomb with a deep crucifix hewn into it. The Archbishop of Canterbury was in attendance. Benedict Cumberbatch, in anticipation of his playing Shakespeare's version in 2016, read a poem by the royally appointed poet laureate Carol Ann Duffy, and though the top brass monarchy were not present, the Queen's daughter-in-law Sophie, Countess of Wessex was, presumably to make sure he wasn't going to make a third comeback and claim the throne back from the lineage of the Tudors. It is a great tale, and unique. Identifying anyone from centuries-old history requires sleuthing skills of the highest caliber. His royalty was essential in establishing living descendants, as those are the people for whom family trees are best known. In doing so, the possible illegitimacy of the reign of all subsequent British monarchs was mooted, but I suspect this challenge will not shake the foundations of Buckingham Palace.

Some of the players in the twenty-first century act of Richard's life still squabble over his character and the evidence, and his role in the shaping of modern Britain. These debates are probably unwinnable. But thanks to DNA, at least we know it is him. Now that crooked usurper king, the bloody dog, lies in his ultimate tomb, inside Leicester Cathedral, and has made his final

Exit.

iii: The king is dead . . .

Madrid; November 1, 1700

. . . But there is no one to succeed him. Charles II, the last member of the imperial Hapsburg dynasty to rule Spain, died five days short of his thirty-ninth birthday. Despite having been married twice, he left behind neither a child nor an heir.

His inability to produce that son who would continue the rule of the House of Hapsburg was just one of a litany of medical troubles Charles suffered during his pitiful life. He was profoundly disabled, epileptic, and mentally incompetent. He didn't learn to walk until he was four, nor talk until he was eight, and when he did, the permanently swollen tongue that filled his mouth rendered his words garbled and his speech dribbly.

Toward the end of his troubled life, it was worse, with his frame crippled and convulsive fits bewitching him. And he became weirder and weirder, plagued with hallucinations, and bizarrely demanding that his ancestors be exhumed so he could gaze upon their decomposing corpses. Even in death, the unusualness of his hexed existence continued, with the postmortem physician describing a battery of internal deformities:

> his heart was the size of a peppercorn; his lungs corroded; his intestines rotten and gangrenous; he had a single testicle, black as coal, and his head was full of water.

Admittedly I'm not a pathologist, but I think that this might be exaggerated. Nevertheless, Charles was unequivocally not a well man at any stage of his divine tenure on Earth, either physically or mentally. As a child monarch, though, he was given special dispensations neither to attend school nor frequently wash. When he was three, his father died, and his mother, Mariana, acted as Regent. She continued even after he reached fourteen—the age at which the Spanish allowed reign single-handed—on the grounds that though physically present, mentally he was *Rex in Absentia*. For the duration of his ineffective rule, Spain was in trouble, and the mad king reeked. To literally add insult, the subjects nominally under his divine supremacy gave him a nickname: *Carlos el Hechizado*—Charles the Hexed, or Bewitched. To die at thirty-nine in the seventeenth century was not a notably short life, but to survive for so long with such profound disabilities was extremely unusual. He endured his agonized years pampered and doted upon, bald yet bewigged, crippled, and epileptic, with the servitude and care that only supreme power can bestow. But it was all for nothing.

His hex left him childless. He was married at eighteen to Marie Louise of Orléans, who died a decade later, and again at twenty-nine to Maria Anna of Neuburg. His first wife spoke privately of his premature ejaculation, and his second of his impotency. These were symptoms alongside a suite of disabilities so profound and numerous that no single disease was or can be posthumously identified. Charles II was the last of the ruling Spanish Hapsburgs, a family of inimitable power and wealth that had delivered every Holy Roman Emperor for 200 years until his death in 1700, and governed the largest chunk of mainland Europe during that time. They bore the inimitable Hapsburg Lip—a prognathous jaw in modern medical parlance—that stuck out literally; a mark of divine power passed from father to son and via many daughters and mothers for generations.

That was precisely the problem. The Hapsburg family's tenacious grip on power meant that they stopped outbreeding more than a century before Charles' birth. His father was Philip IV of Spain (a proud Hapsburg chin is on prominent display in his portrait in the Dulwich Picture Gallery in southeast London) and his mother Mariana of Austria. Her mother was Maria Anna of Spain, who was Philip's sister (see page 198). Therefore, Charles' maternal grandmother was his paternal aunt. Philip and Maria Anna's mother was Margarita of Austria, which means that she was Charles' grandmother and great-grandmother at the same time. On the pedigree itself, these are clearly visible as closed loops, when they should be open branches. In another loop, Charles II of Austria was both great-grandfather and great-great-grandfather. These marriages were frequently intergenerational, often an uncle marrying a niece, so Anna of Hapsburg (1528–90) was Charles' great-great-aunt, great-great-great-aunt, great-great-great-grandmother, and great-great-great-great-grandmother twice. Six generations should contain 62 different people, as does mine, and our own queen's today. Charles II's has 32. Eight generations should contain 254 different people. Charles II's has 82.

Indulge me one more, because this whole dynasty is set up by the marriage of Philip the Handsome and Joanna of Castille in 1496. The shortest route from Charles to Joanna is five generations, which occurs twice, but she is also six generations from him via five routes, and seven twice. Over those three generations, fifth, sixth, and seventh, you should in principle have 112 women. Joanna of Castille occupied nine of those positions on her own across three generations. This is not desirable.

Joanna had a nickname too. In her later years she suffered from some form of mental illness, and modern speculative diagnoses have included melancholia, depression, psychosis, and schizophrenia. Some historians have suggested it might have been inherited:

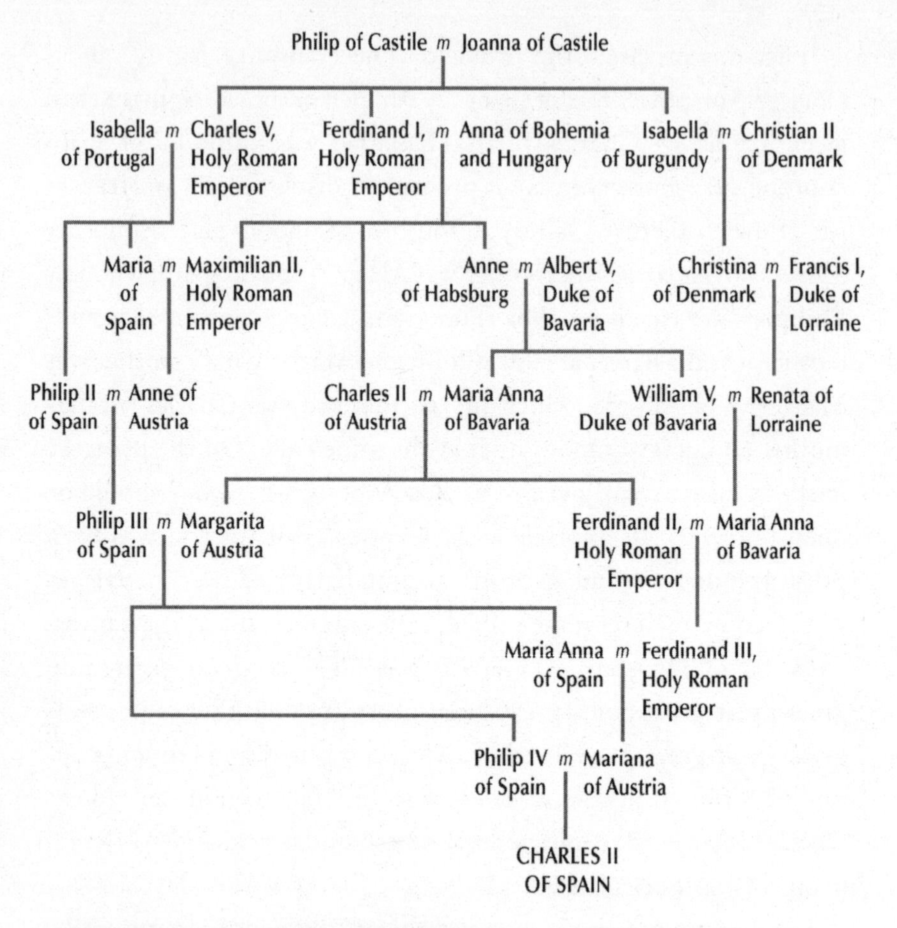

The family tree of Charles II of Spain.
The Hapsburg family stopped outbreeding more than a century before Charles II's birth in 1661. Six generations should contain 62 different ancestors; Charles II's has 32. Eight generations should contain 254 different ancestors; Charles II's has 82. This is what we might refer to as "suboptimal."

Her grandmother, Isabella of Portugal, also suffered in a similar way. This type of posthumous medicine is always difficult to pin down without physical remains or DNA, and supposition abounds. But at the time, her subjects were not that fussed about a clinically precise diagnosis: They called her *Juana la Loca*—Joanna the Mad.

As examined in the previous sections, family trees are much less arboreal than we might assume, and on a long enough timeline we're all inbred. One hopes, though, that your immediate family branches outward and upward without a loop in sight. The Hapsburg tree that ends in Charles II is a tragic thicket, and doomed to what is known as "pedigree collapse."

Actually, all our pedigrees will collapse in time. The story of Charlemagne earlier in this chapter is a description of every European's pedigree collapsing onto everyone in medieval Europe. But Charles' collapses in on itself over a painfully shorter timescale. Inbreeding in its simplest terms is the failure to add genetic variation to a child because of mating with someone closely related. You share half your genes with each of your parents, and therefore a quarter with each of your grandparents. Distasteful (as well as illegal and immoral) though it is, if a father and daughter (or mother and son) were to reproduce, their child would still have four grandparents, but one of them would also be a parent, and so, that person would not be contributing any new genes into the mix. A child of two half siblings has only three grandparents. The children of two first cousins will have four grandparents but only six great-grandparents.

The closest inbreeding that can be managed in a single pairing is between a brother and sister—this is a maximum pedigree collapse in a single generation. Any closer and they'd be clones (non-identical twins are genetically equivalent to brother and sister, and identical twins can't have reproductive sex, obviously). They have exactly the same parents and grandparents, and so their child's genes are drawn from the same pools and no new genes will be contributed for at least two generations. We each have two copies of every gene, one allele—as the variants are known—on each chromosome, one from each parent. This system is an insurance policy, and exists to ensure that a good one hopefully masks a bum allele.

For the child of a brother and sister, one quarter of their total genomes, on average, of all their genes will be the same on both of the pairs of chromosomes, one to twenty-three. The chances that some of these will be alleles that are recessively dangerous is high, and in an outbred child, they would be masked by the influx of different versions. The measure is called an inbreeding coefficient, given the letter F,* and the child of a brother and sister will have an F value of 0.25.

In 2009, Gonzalo Alvarez and his team of Spanish geneticists went back sixteen generations from Charles II, and incorporated more than 3,000 individuals to calculate his F value. It came out as 0.254. Charles, through generation after generation of uncle-niece pairings or first cousin pairings, was more cumulatively inbred than the child of a brother and sister. Though the family tree is an obvious disaster, this was the first time the collapse of the Hapsburgs had been considered from a genetic point of view. Philip I was calculated to be a relatively normal (though still high) 0.02, so in the 200 years between him and Charles II, according to the standard measure, their family had become ten times more inbred. Philip III was also a high scorer, F = 0.218, as he was, like Charles, the son of uncle-niece sex. Alvarez showed that the blood ran deep too, well beyond the five or six generations that start with Philip and Mad Joanna. When adding the levels of inbreeding each generation up, it only begins to level off for Charles II and Philip III after ten generations, which means that less obvious levels of inbreeding were also slowly trickling into this doomed lineage hundreds of years earlier.

---

* Related to "fitness," not for whatever you might be thinking about sex. Fitness is the key measure in evolution, and quantifies reproductive success—that is, how many offspring, on average, an organism will have in a population, or more precisely, the contribution that an individual makes to the subsequent generation gene pool.

We also see a tragic death rate among the children of this clan. The family tree, while almost comical in its perversion, does not show the mortality of the young. Between 1527 (when Philip II was born) and Charles' birth in 1661, the Hapsburgs had thirty-four babies. Seventeen of them died before they were ten, and ten of them before they were one year old. The infant mortality rate in Spanish subjects—including peasants, with not nearly the medical treatment or privilege of the wealthy—was around one in five at that time.

This had a cumulative effect on the dynasty. All hope of succession was put upon Charles, with his shriveled genitals, because so many of the other potential heirs had died in infancy. It's not possible to guess what their medical problems might have been, nor indeed is it easy for Charles, such were the profound shallows of his genes. Of all of Charles' DNA, more than one quarter was identical in both copies. Having one copy of a recessive disease gene will be masked by a normal copy; two copies means the disease is revealed. In his case, dozens of recessive problems were exposed, rather than hidden. Alvarez suggests two disorders that might have emerged from his DNA. PROP1 is a gene that is active in the pituitary gland, a pea-like part of the brain at its base. It's an organ that produces an array of hormones essential for growth and a host of other normal bodily functions. Combined pituitary hormone deficiency occurs when you have two copies of a mutated PROP1, which includes a panoply of symptoms, including feeble muscles, infertility, impotence, and a strange apathy about one's surroundings. They propose a second recessive disease, distal renal tubular acidosis, which is caused by mutations in two genes, and includes weak muscles, rickets, and a large head in proportion to one's body.

That's all a good fit for the descriptions of Charles' hexed life, but Alvarez stresses that these are speculations. To be quite that inbred may open the door to specific recessive conditions such as these, but there's dozens of other things it could be, and the overall genetic robustness—and therefore actual health—of such an individual would be horrifically chaotic. A seventeenth-century grasp of genetics ironically undid their pathological lust for power. The tighter they squeezed their grip on the reins of power, the sooner the House of Hapsburg would fall. He was cursed by his parents, and their parents, and uncles and aunts, his life bewitched by greed. Charles II of Spain—Carlos el Hechizado—is an extreme example of what happens when you keep it in the family.

The importance of not being inbred

Inbreeding is an essential component of science. Genetics would be nowhere without it. Mendel's pea crosses were cleverly designed backcrosses so that the same trait would be forced back into generations so he could calculate the proportions of each trait. Mice are the human geneticist's first choice model organism (with a big nod to the fruit fly, and also the rat, nematode worm, and zebrafish). That's only partly by design. The first mouse genetics experimenters got their animals from competitive breeders at the end of the nineteenth century in America. These mouse fanciers competed in shows, just like Westminster and other dog shows today, and strove to produce mice with ever more colorful coats. In order to keep competing year after year, the best ones would be studded to retain the best traits.

As anyone who has dogs will know, pedigrees have all sorts of problems, again as a result of inbreeding. My family has black flat-coated retrievers, and our first had a congenital heart defect, a murmur, which she endured through a thirteen-year life without symptoms. Her siblings were not so lucky and had a raft of health troubles. To this day, enormous effort goes into making sure that the genetics of the mice we use in every experiment are inbred enough to reduce any natural variation that might skew the results, but not so inbred that they are diseased.

Darwin worried about the fact that his beloved wife Emma was his first cousin. He fretted that this proximity would impose unnecessary health problems. They had ten children in all, but three of them died before the age of eleven. He fretted about the health of the whole family, writing:

> it is the great drawback to my happiness, that they are not very robust; some of them seem to have inherited my detestable constitution.

Mind you, Darwin fretted about a lot of stuff, especially his health, his kids, and maybe with just cause. On occasion he would write a fit of histrionic despair, such as "I am very poorly today & very stupid & hate everybody & everything," and once, evoking the spirit of eugenics soon to be invented by his cousin, Francis Galton, about whom we will hear more later:

> We are a wretched family & ought to be exterminated.

Their health was filled with tragedy. Annie (number 2), arguably his favorite, died aged ten, possibly of tuberculosis, after a bout of scarlet fever. Mary Eleanor (number 3) died just twenty-three days after her birth. Horace (number 9) arrived shortly after Annie's death, but had "attacks, many times a day, of shuddering & gasping & hysterical sobbing, semi-convulsive movements." Charles thought Leonard (number 8) "rather slow and backward,"* though he lived to be ninety-three, was a Member of Parliament, and his

---

* Clearly a family trait: Charles' father Robert Darwin once wrote to his son aged sixteen: "You care for nothing but shooting, dogs and rat-catching and you will be a disgrace to yourself and all your family." It is unknown whether he later said, "Son, sorry about that, seeing as you turned out to be pretty much the greatest Englishman ever. Top ten, easily."

scientific legacy is impressive: He went on the missions to study the Transit of Venus in 1874 and 1882, when that planet passes in between us and the Sun, and from which we learn magnificent amounts about planets. And in the twentieth century Leonard was patron to Sir Ronald Fisher, one of the great evolutionary biologists, who in the 1930s helped fuse the new genetics with natural selection, and set up the framework for the future of genetics. Leonard was the second president of the British Eugenics Society, succeeding Francis Galton.

Darwin's anxieties came not just from mild hypochondria, but from the testicles of the plant world. Orchids are named after the shape of their roots,* which resemble the male gonads, called *orchis* in Greek. He bred them by the dozen, fifty-seven species in all, and noted and concluded that cross-fertilization resulted in more robust plants than self-pollination. Those that had been crossed with themselves tended to be shorter, less massive, later flowering, and produced fewer seeds. This phenomenon is known as an inbreeding depression.

Nothing like the mire of the Hapsburgs, but the Darwin-Wedgwood clan was prone to a spot of self-pollination. The geneticist Gonzalo Alvarez had climbed the Hapsburg family tree in 2009, and turned his attention to the Darwins in 2015. Charles and Emma were first cousins, as indeed were the marriages of three of her brothers. This gives their offspring an F value of 0.0625. Emma's brother Henry Wedgwood and his wife Jessie Wedgwood were double first cousins—both their fathers were brothers and both their mothers were sisters, and their children had F values of 0.1255.

Then again, three of Emma's siblings married out, and if, as with the Spanish Hapsburgs, we use infant mortality as a measure

* The Middle English name for orchid is bollockwort. This deserves to be reintroduced to polite contemporary society.

of genetic robustness, the Darwins do OK. Over four generations, 25 individual Darwin-Wedgwood families had 176 children, 155 of whom lived. This is a child mortality rate of about 12 percent, which is less than the national level in midcentury Victorian England of about 15 percent. Then again, they were a rich, privileged pedigree, and enjoyed all the health and medicine that upper-class Victorians could afford.

Inbreeding has been studied in depth across the living world, notably in plants and in conservation efforts, where depression can point a species in the direction of extinction. In humans it's been studied on and off throughout the twentieth century and into the twenty-first. Results have been mixed. In the 1960s, studies on Japanese islanders showed no great measurable relationship between inbreeding and overall health. In the 1970s, intellectual disability was shown to be more likely in consanguineous marriages. And in 2014, a strong correlation was shown between inbreeding and low performance on IQ tests in Muslim families in north India. Intelligence and its inheritance is a complex business and requires a whole book in itself to delve into that world. IQ is a measure of an aspect of broader cognitive abilities, and it gets bandied about all the time. Often it's asserted that IQ only measures how good you are at IQ tests, which is glibly true, in the same way that running the hundred meters only really tests how good you are at running as fast as you can over a hundred meters. Regardless of the controversies that follow the study of intelligence, IQ remains probably the best-researched aspect of cognitive ability, and that, at least, gives it some value for scientists.

But just to simplify matters, a massive study of the effects of inbreeding was enabled by the new genetics, and conducted in 2012, on a characteristic that is much better understood: height. We understand height well. We know its genetics well, even though it is influenced by many factors, both genetic and environmental (the concept of heritability is discussed in Chapter 6). Edinburgh geneticist Ruth McQuillan and a massive international team scanned the DNA of 35,000 people across twenty-one European countries, and looked for identical genetic variants on both chromosomes across the whole genomes, thereby identifying consanguinity. The difference between those with first-cousin-like inbreeding and outbred individuals was, on average, three centimeters. Socioeconomic factors played a massive part, and were accounted for in the final analysis. The inbreeding depression on one of the simpler metrics of humans is just over an inch.

The black Tinkler winna get ye

It is easy to overstate the impact of inbreeding. In the UK, first cousin marriages became a political hot potato in the second decade of the twenty-first century when MPs began commenting in public on rates of first cousin marriages in Pakistanis, in Muslims, and in the city of Bradford. These are three overlapping groups. There are around 1 million Britons of Pakistani origin and around 70 percent of them come from the Mirpur region in Pakistan. Some studies have shown that 60 percent of the Mirpur Pakistani marriages in Britain are between first cousins. These numbers are a bit hazy, but by my reckoning that makes around 200,000 first cousin marriages, which is rather high. Not all Muslims endorse first cousin marriage, but perhaps it is fair to say that Islam is more accepting of that pairing than other cultures or religions, though Buddhists and Zoroastrians are also pretty relaxed about it. What are the risks?

A first cousin marriage almost doubles the risk of a recessive disease emerging, compared to non-consanguineous parents. But the numbers are low to start with. The probability overall for non-related couples having a child bearing a recessive disease is around 2 to 3 percent. For first cousins it's around 5 percent. This, by the way, is about the same as for a couple where the woman is over forty-one. So, there is a risk, but it's not huge, and perhaps some medically informed discussion about the risks is in order, especially among communities in which first cousin marriages are sanctioned or even encouraged. Certainly, stigmatizing them does not seem like a constructive move. Politicians seem all too willing in the UK to make some pronouncement on the medical risks of first cousin marriages. In 2008, the now disgraced* Labour MP for Oldham East and Saddleworth, Phil Woolas, who later became the minister for immigration and borders, said:

> If you have a child with your cousin the likelihood is there'll be a genetic problem.

This is true in a heavily qualified sense, but offers little insight into the real risks. Some research had shown in 2005 that British Pakistanis, while generating 3.4 percent of the children born in the UK, accounted for 30 percent of the recessive congenital disorders. That is a real issue, but genetic counseling is a real profession, where experts advise prospective parents on their risks of genetic diseases emerging in their kids; education about the risks of consanguineous births can be presented to parents without stigmatization. Many Muslims took Woolas' comments, and indeed the equally clumsy words of MP Ann Cryer in 2006, and even the heavily nuanced and scientifically accurate words of Steve Jones (professor of genetics; my tutor, to whom this book is dedicated) as attacks on Muslims. Jones' words were balanced by his comparisons to similar levels of inbreeding in Northern Ireland, and the

* Woolas was removed from his seat in Parliament in 2010 for allegations about an opponent during the election campaign, and was thrown out of the Labour Party.

British royal family, though predictably these qualifications were far less reported. In fact, in many Islamic cultures, imams are trained to offer genetic counseling about the risks of recessive diseases emerging as a result of consanguineous marriages. This is not ubiquitous in Islam, not least because it is a broad church, with many different practices.

The reverse stigma also has raised its ugly head in the recent past, in the form of casual prejudice against Roma. In October 2013, the Irish police seized two children from their Roma families and homes. In Greece, a few days earlier, the police raided a home and removed a five-year-old girl from her parents. All three of those children were blond and blue or green eyed. The abduction of Madeleine McCann dominated the press for several months in 2007, and has never truly gone away, as the search for her continues. References were made in the press to the McCann case as these other children were seized under veils of suspicion, drawn partially from the incongruity of blond children being in the presence of dark-skinned parents. Much of that prejudice runs centuries deep. *The Adventures of Dollie*—the debut film of that founder of American cinema D.W. Griffith—is a single reel tale of a "gypsy" kidnapping a white girl made in 1908. There are dozens of documented cases of Roma or Irish Travelers being falsely accused of stealing children over the twentieth century, and back through modern European history, to the extent that it features in a nineteenth-century nursery rhyme:

> Hush nae, hush nae, dinna fret ye,
> The black Tinkler winna get ye.

According to Thomas Acton, a professor of Romani studies at the University of Greenwich, there isn't a single verifiable case of Roma stealing non-Roma children in history.

In the modern age, it's the clash of genetics and prejudice that fuels the myth. How could such a "blonde angel," as Maria, the Greek girl, was rechristened by the press, not be a changeling, whisked away by swarthy Gypsies that were raising her? In fact, the two Irish children were returned to their homes following DNA tests that proved that they were being raised by their biological parents. Oddly, DNA revealed that the dark-skinned people raising Maria in Greece were not her birth parents, but after further enquiries her real Roma parents were found in Bulgaria. They had informally given Maria up for adoption to the Greek foster parents owing to their own poverty.

This blondness was not news to any geneticist. California-based genetics professor Leonid Kruglyak told me he was surprised anyone thought these kids *couldn't* be blond. Nevertheless, the assumption that it was a genetic impossibility helped fuel suspicion because it tied neatly into a narrative, an ancient prejudice that Gypsies steal non-Roma babies. The Roma are a community of travelers whose genetics is somewhat peculiar. Estimates vary, but there are between 8 and 11 million Roma worldwide, most living in Europe. They emerged from India probably around the year 500 CE, according to a genetic analysis published in 2012, migrated through to the Balkans, and dispersed via Bulgaria around the eleventh century. There are Roma (also called Romani) enclaves in many European countries, and they show relatively high degrees of endogamy—mating within a social or cultural group—but also some breeding with local indigenous people. Nevertheless, the population as a whole appears to have had at least two bottlenecks, initially after leaving India, and then later as the Roma began to disperse into their different populations.

Estimates about levels of inbreeding are difficult to come by for the various Roma communities, but studies have tried to establish some of the underlying genetics of these people. More than nine different mutations that appear to be unique to Roma communities have been discovered, and a high level of carrier status for a few more. One of these is oculocutaneous albinism, a recessive condition that results in a lack of or highly reduced pigmentation in people with two mutated genes. There are different types, but people with OCA1B are not devoid of color as you think of typical albinos. Instead they have very light blond hair at birth, that darkens a touch over time, pale skin that slightly tans, and blue or green/hazel eyes. In Spanish Roma, the prevalence of carrier status for OCA1 is around 3.4 percent, possibly seven times higher than in the non-Roma population. With this high rate, at least in this Spanish population, which is a small and largely endogamous community, the presence of blond children in Roma families is virtually a genetic inevitability. After Maria's Bulgarian birth parents were found, it was reported that of their nine children, five of them are blond and blue eyed, as are two of their paternal uncles.

Were the Darwins in an inbreeding depression? It's difficult to say one way or another. The Hapsburgs plumbed unprecedented depths into a hole of consanguinity from which they could not climb out. But large Victorian families often married fairly close to home, and although childhood mortality in their clan was high, over the years the Darwin-Wedgwood-Keynes boast a fecund and impressive family tree. First cousin marriages were made legal in Britain by Henry VIII in the sixteenth century. In the West only the USA has them banned, in thirty-one states.

In the present day, in the cultures in which first cousin marriages are permitted, and even encouraged, we do see broadly higher rates of disease, and signs of physical well-being that are reduced as well. But it's not a huge increase. Roma communities do display higher levels of recessive disorders than the background population, but they nevertheless produce large clans. Shock headlines and mock outrage by concerned politicians aren't particularly useful.

Then again, we know about so many diseases and fundamentals of genetics by the undesigned experiments humans indulge in when they marry their kin. The identification of Tay-Sachs disease was enabled by its presence in consanguineous families in the late nineteenth century. Our understanding of the genetics of speech was deepened when an inbred family in Great Ormond Street Hospital presented a range of vowel and consonant problems, and a faulty version of the gene FOXP2 on chromosome 7 was found as the cause. I worked on a form of blindness called microphthalmia, and our identification of the gene that caused it was only possible as it was seen in inbred Middle Eastern families. The list of genes that have been found because of consanguinity is long.

In December 2015, the UK's 1000 Genomes project began winding up its detailed survey of not its titular number, but more than 2,500 complete human sequences from twenty-six different peoples. They hadn't snooped around in this data for signs of close family relationships, but they certainly found them, in all twenty-six populations. A quarter of the individuals in the sample were classified as inbred. One in twenty-five were as inbred as first cousins, and in some populations the number was half. Two hundred and twenty-seven of them were found to be first cousins (or the genetic equivalent) that were unknown (or at least unreported). Fifteen were closer than first cousin; eight between parent and offspring relationships, three brother and sister, one half sibling, three uncle-niece relationships.

The project is not designed to expose consanguineous relations nor shame them; the individuals remain anonymous throughout. Instead these inbreeding calculations were made by looking at blocks in the genome that are the same in both chromosomes and the same in their parents—homozygous by descent. The spread was not even in geography or culture either. South Asians were the most inbred, on average, meaning people from the Indian subcontinent. Africans and Europeans were the most outbred, though among the Europeans, the Finns came at the out top of the inbreeding tables. That again is not surprising, as that is a country with a small founding population and not a lot of immigration. This is of great importance to further studies on how the genome works, and the patterns of human variation around the planet. Two thousand genomes are now a database for researchers to scour, and understanding the level of inbreeding in those people is essential for us to recognize the effect of consanguinity on our understanding of genetics. Accounting for the relatedness of participants in studies is invaluable.

We are perversely indebted to the inbred; the Icelanders, Jews, Finns, Persians, Indians, Pakistanis, in fact all populations, not forgetting the pigeons, mice, fruit flies, and worms in which consanguinity has revealed the workings of our genomes. There is an increased risk of the emergence of recessive genetic diseases when new genes fail to make their way into the genomes of children. But it's easy to overstate them, and easier to allow cultural practices to foster and foment prejudices. Education and genetic counseling is available to all, at least in the UK, and we would do well to relieve the burden of genetic diseases on families and on healthcare by encouraging those for whom consanguinity is normal to embrace these services. But, as long as people continue to do it, geneticists will look to the genomes of the closely related families to see what particular secrets they hold over the outbred. Nevertheless, no matter how extremely interesting inbreeding is for geneticists, if you keep doing it generation after generation, it's not good news for families, royal or otherwise.

PART TWO

Who We Are Now

5
The end of race

"I think I shall avoid the whole subject as so surrounded with prejudices"

Letter from Charles Darwin to Alfred Russel Wallace, December 22, 1887

October 1981, Capel St. Mary, Suffolk

I was Leroy and my sister was Coco when I first encountered racism. We were at the Co-op supermarket in the tiny village of Capel St. Mary in Suffolk, where we lived, when some boys on their bikes shouted out those names to my eight-year-old sister and me, seventeen months younger. *Fame* was the big hit on television at the time, and Coco and Leroy were the lead characters. I remember two things about what happened next. First, we thought this was excellent, as they were amazing dancers, both African Americans, both supremely attractive: Leroy surly and charismatic with cornrow hair, and Coco tough and beautiful.

The second thing I recall was that my father was livid, splenetic with rage. Maybe his reaction was piqued by the recent murder of his best friend, Blair Peach, by a policeman at an Anti-Nazi League rally in London a year or so earlier. At the time we were utterly baffled by Dad's reaction, especially as Coco and Leroy were so effortlessly cool. I guess being slightly dark-skinned in rural Suffolk in the 1980s simply made us stand out. My memory is that my sister and I did bad splits all the way home.

Not long after, in the first term at a new school, a fellow seven-year-old called me a "Paki." By this stage I knew that this was intended as an insult, and I punched him, I think in the stomach.* Like most bullies do when confronted, he went crying off to teacher and I was summoned to the headmaster's office. With an understandable zero tolerance policy on violence, Mr. Yelland was spittle furious and I was terrified. I told him what happened, he bellowed for me to leave, which I did, still petrified. It seems he had a less-than-zero tolerance level for racism. The other boy got a week of detention.

These two trivial examples of fairly low-level racism are pretty insignificant, and I can't pretend that I have endured much racial prejudice in my life. My mother is Indian, but I am not very dark, and I sometimes am assumed to be Italian. I have a robustly British name, and the type of accent that gets you a job as a presenter on BBC Radio 4. I have visited India several times, but only as a tourist or for work. I did find my roots of sorts on one trip, but it was of limited scientific interest, revealing that my maternal DNA was likely to have arisen somewhere in the region of Mumbai, probably some 20,000 years ago. There was certainly no spiritual epiphany—I am as Indian as cricket. My mother has never set foot on the subcontinent, nor had her parents. Before they were born, her grandparents immigrated to Guyana in South America, where my mother spent the first twenty years of her life before moving to England, as so many Guyanese women did, when Britain looked to the embers of the colonies to recruit nurses for the neonatal National Health Service. She now lives in Canada.

My father was born in Scarborough, in Yorkshire, and his family tree is rooted in the northeast of England, and north of the border as far back as we can trace, to the seventeenth century. The name

* Punching racists is not necessarily something that I would encourage or endorse.

Rutherford strongly associates with those areas, and it has its own tartan and motto: *nec sorte, nec fato*—neither by chance, nor fate—which I suppose is agreeably relevant to evolution and genetics.

His family emigrated to New Zealand when he was five, where there have been plenty of other Rutherford settlers, not least the great physicist Ernest,* but he was drawn back home at twenty. His sister, my aunt, still lives on the other side of Earth and, with an Austrian Jew, had a son, my cousin, who is a full Kiwi, as are his two daughters, my first cousins once removed. My parents divorced when I was about eight, and we soon moved in with my father and his new partner, now wife, who raised us as her own. She is East Anglian, and we can see her maternal family tombstones in a single Essex churchyard all the way back to the seventeenth century. Her father and aunt were orphans raised by Sisters of Mercy in Liverpool, though when we investigated their family tree it looks like their father, her grandfather, was a Russian Jew named Josef Abrahams, but he took (or was given) the naturalized name Joseph Adams—so unceremoniously is heritage abandoned.† She already had two sons, and fused our lineages when together she and my father had another, my youngest brother, sixteen years my junior.

We're all adults now. My sister married an English South African with Dutch and German/Jewish ancestry, and they have two daughters. My middle brother married a Swede and lives in the

* No relation.

† This is not uncommon among Jews. I have a friend who is a Cohen, but not because of a long tradition of priesthood as is typically the case. It is merely that upon arrival in the UK in the nineteenth century from eastern Europe, unable to speak English to immigration officials, their Jewishness was established, and Cohen was assigned as a typically Jewish name. Similarly, I have friends with the surnames Gee and Kay, because having an obviously Jewish name like Ginsberg in the 1930s was potentially dangerous, and names such as Krupnik were perceived as being difficult. Both were simply shortened to the first letter.

birthplace of biological taxonomy, Uppsala,* where they have two dual-nationality bilingual kids. My youngest brother lives with an Iranian woman whose family is in exile. My stepmum's sister married a man who was adopted by an English family after being found as a baby abandoned in Greece. They have three boys; the eldest married an English woman with a half-German half-Maltese mother. I married an English girl, with Irish and Welsh ancestry on either side of her parents. Our three kids were all delivered by me in London.†

So it goes. In one generation, a Russian Jewish family becomes British Gentile, my cousin is born a New Zealander, and my niece and nephew Swedish. Sometimes, when asked by people or on pull-down menus on official forms, I describe myself as mixed-race Indian British, but is it true? I'm from Ipswich, though now, as I enter my fifth decade, I've spent more time in London. I think of my pedigree as mutt—less a tree than an amusing tangled thicket.

Are we unusual? Though through history people have typically married close to home, the progress toward out-breeding has been mostly relentless. At least, this is true in un-royal families. And we've seen (in Chapter 4) how inbreeding worked out for the Hapsburgs—that is, not well. It's no great revelation to point out that racial abuse is idiotic, but being called a "Paki" doesn't make a great deal of sense given my background. It's even more absurd to compare my sister and me to African Americans, no matter how awesomely cool Coco and Leroy were, or we might've hoped to be.‡ Without knowing the ancestry of the actors, Irene Cara and Gene

---

* Where Carl Linneaus lived and died, having reinvented biology and specifically the system by which we classify all living things.

† I believe the geography of my son's birth qualifies him as a Cockney.

‡ Becoming a scientist and a museum curator, respectively, rather than dancers surely answers this pressing question.

Anthony Ray, it's quite possible, extremely probable in fact, that I am more closely related genetically to the racist boys in Suffolk, than those two black kids from *Fame* are to each other. This applies to both my paternal, that is British, genes, and my Indian maternal ancestry.

In essence, this is because of the small size of the populations that slowly ambled out of Africa some 100,000 years ago, growing and migrating to populate the rest of the world, as discussed in Chapter 1. Only a relative few formed the genetic pool from which we were drawn, far smaller than the people who remained in the nursery of humankind. In statistics this is called a sampling error. By drawing just a small taster from a big population, you select a sample that is not representative of the original group. That small misrepresentative sample was the group of people from whom all DNA would be drawn, apart from those in Africa. This means that in fact, I would be genetically more similar to those Suffolk kids if I were Russian, Swedish, or Maori than many people of largely African descent are to each other.

I guess they didn't have a working knowledge of human genetic variation and migration, and I feel compelled to conclude that they weren't intending to make a comment on the evolutionary ascent of man. Racism is hateful bullying, and a means of reinforcing self-identity at the expense of others: Whatever you are, you're not one of us. If there is one thing that my own tortuous family tree demonstrates it is that families make a mockery of racial epithets, and racial definitions as used in common parlance are deeply problematic. Modern genetics has shown just that, too, and I'll be navigating through some of the data on that in the next few pages. But here is the idea I will be interrogating: There are no essential genetic elements for any particular group of people who might be identified as a "race." As far as genetics is concerned, race does not exist.

I am unaware of any group of people on Earth that can be defined by their DNA in a scientifically satisfactory way. There are plenty of genetic differences and physical differences that emerge from those genes between people and peoples, but none that align with the way we talk about "race." The question of what race means from a scientific point of view is complex, controversial, and still a source of great ire and debate. It is frequently stated that, for the average geneticist, race simply does not exist. This chapter will explore how true that is, not for political reasons, though it is virtually impossible not to feel a sense of right or wrong or injustice or moral indignation when we talk about race. Science's primary role is to subtract those human characteristics from objective reality, and cast a picture of how things are, rather than how we see them.

The great irony is this: The science of genetics was founded specifically on the study of racial inequality, by a racist. The history of my field is inextricably intertwined with ideas that we now find toxic: racism, empire, prejudice, and eugenics. Like all true stories and histories, like all real family trees, what follows also weaves a meandering path. All geneticists, all statisticians,* and in fact all scientists owe this Victorian racist a profound intellectual debt, for he was a genius on whose foundations much of the modern world rests. His name was Francis Galton, and as all the great stories in biology do, it begins with Charles Darwin.

---

* Modern genetics is inextricably enmeshed with statistics. Evolutionary biology enjoyed a giant boost in the first half of the twentieth century when math and statistics were applied to Darwin's ideas of natural selection by some of the finest scientists ever to have drawn breath, many of whom held deeply unpleasant views—the names Pearson and Spearman will be familiar to anyone who has dallied with statistics in their lives. In the era of genomics, statistics plays an even more pivotal role in uncovering the relationship of human populations to each other, but nowadays most statisticians are not appalling racists.

Darwin had published *On the Origin of Species** in November 1859, to immediate blockbuster success. The first edition sold well immediately (though it is unknown whether it sold out, as Darwin wrote in his diaries, and is often repeated), and Darwin amended his work and a second edition of 3,000 copies followed in January 1860. He wasn't a terribly well man, possibly having picked up a disease (still precisely undiagnosed today) on his travels on HMS *Beagle*, where he saw evidence of evolution and of the grinding movements of land on the planet, and began to think about the spread of life on Earth, and the similarities and differences between the peoples of the continents. Though voluminous in his correspondence, Darwin didn't venture out in public often. But his defenders, notably Thomas Huxley and Joseph Hooker, were vociferous in their support for his central idea of evolution by natural selection, descent with modification. By the 1850s, Darwin was already well known from his global travels, but on publication of the *Origin*, he became a star. The word *genius* gets bandied around rather casually these days, but Darwin was one unequivocally by any definition—in my view, the greatest of all scientists, across all disciplines. No one singly has done more to reposition life and us in the universe. He described why life on Earth is the way it is, and placed humans very clearly in relation to other animals. Furthermore, though normally I find ranking crude and idolatry disappointing, I'm more than happy to

* The full title (of the sixth edition sitting on a shelf to my right; other editions vary) is *The Origin of Species by Means of Natural Selection or the Preservation of Favoured Races in the Struggle for Life*. Note the word *races*; in this contemporary context the word is used in a way that we no longer accept, meaning something akin to breeds, or subspecies or simply types of organism. I point this out because superficial critics of both far-right and antiracist sensibilities have attempted in public and private correspondence to use this as some sort of criticism or propaganda for their causes. If they had made it past the title, and all the way to the final chapter, they would know that Darwin mentions humans just once in that great book, quoted as the epigram at the very beginning of this book, and there only to ponder what his theory means for us bald apes.

single him out at the very top of the intellectual pile because he was so humble, and credited every Tom, Dick, and Harriet who had the slightest influence on his scheme.

Francis Galton was a genius too, but also a difficult man, and a man more difficult to scrutinize. He was Darwin's half cousin, and profoundly influenced by the runaway success of Darwin's work and ideas. *Half cousin* means that they shared a grandfather, Erasmus Darwin, who married twice. Erasmus was a great thinker and scientist too, and alongside Samuel Galton a founder member of the magnificent Lunar Society, a Birmingham-based intellectual salon of industrialists, such as Josiah Wedgwood and James Watt, and scientists, who would meet by moonlight (to aid potentially drink-wobbly walks home late into the night), and scheme up ideas that would come to stoke the white heat of Victorian ingenuity.

Both branches of Erasmus Darwin's family were established, wealthy, and successful. The Darwins were scientists and physicians, and the Galtons were Quakers and gunsmiths, which might seem an unlikely combination given that religion's commitment to nonviolence. Francis was born to Samuel Galton and Frances Darwin in 1822, in a house near Birmingham that was previously occupied by another scientific goliath, the philosopher, theologian, and chemist Joseph Priestley.

Galton had always hero-worshipped his cousin, and was greatly impressed with Darwin's *magnum opus* in 1859. It would have a profound impact on his work. Galton wrote warmly within three weeks of publication to offer characteristically Victorian praise:

> Pray let me add a word of congratulation on the completion of your wonderful volume, to those which I am sure you will have received from every side. I have laid it down in the full enjoyment of a feeling that one rarely experiences after boyish days, of having been initiated into an entirely new province of knowledge, which, nevertheless, connects itself with

> other things in a thousand ways. I hear you are engaged on a second edition.
>
> There is a trivial error in page 68, about rhinoceroses . . .

Aside from his rhino blunder, Darwin's first chapter speaks of breeding in pigeons and plants, and all manner of beasts of the field and fowl of the air. He was demonstrating the mutability of creatures, that they were not fixed in time, but their bodies and behavior could be bent to man's will through selective breeding. This formed the groundwork from which he would go on to describe the natural process of selection.

After reading his cousin's masterwork, Galton began pondering whether humankind could be improved by selective breeding. Darwin was a focused scientist compared to Galton, though that title did not exist until 1834. The somewhat arbitrary subject areas of science that we cling to in school today were not so rigid back then, and most dabbled in multiple fields. Darwin was preoccupied with other living things as well as his pigeons, particularly worms, carnivorous plants, and barnacles,* though he was also driven by geology, which was critical to the development of his evolutionary thinking. Galton, by comparison, was more a polymath, and made not insignificant contributions to a whole range of fields. His myriad gifts to the world included the first newspaper weather map,† the scientific basis of fingerprint analysis for forensics, a dizzying

---

* Darwin's interest in aquatic crustaceans was such that one of his sons reportedly expressed puzzlement that all fathers did not have such an extensive barnacle collection, asking a friend, "Where does your father do his barnacles?" This has been attributed to Leonard, George, and Francis Darwin, making me think that maybe none of them said it. The Darwin Correspondence Project curates his voluminous writings, and was unable to confirm a specific son, so I think it falls into the very well-stocked category of Darwin apocrypha.

† In fact, it was the previous day's weather, in the *Times*, April 1, 1875, and not

number of statistical techniques, many the underpinnings of all statistics used today, foundational work on the psychology of synesthesia, a vented hat to help cool the head while thinking hard,* and much else over his long and distinguished career. He also gave us the word *eugenics*, more of which later, and the phrase *nature versus nurture*, which has plagued geneticists ever since, as this whole book I hope makes amply clear. He devised a new way to cut cakes, which was published in *Nature*, the journal from which both the structure of DNA and the first human genome would break out of the lab and enter the public consciousness.†

My own story, academically speaking, is somewhat enmeshed with Galton's, not in a particularly special way, merely as it is for many whose intellectual genealogy is rooted at University College London. Like both Darwin and Galton (though this is surely where the comparison ends), I went up to university to study medicine. But I wasn't that interested in being a doctor, and spent a lot of time gadding around trying to find something else that would gun my engines. Darwin was similarly distracted from his medical studies in Edinburgh by taxidermy and was rather taken by his tutor John Edmonstone, who was a freed black slave; this relationship was part of Darwin's development as an abolitionist.‡ Instead of lectures, I went to the movies a lot. Galton did a couple of years

---

a joke.

* You could of course simply take the hat off if your head is hot. But these were Victorian times, and I presume a hatless man would be considered unacceptable, immoral, and possibly criminal.

† Which, purely by coincidence, is where I worked for more than a decade.

‡ Darwin trivia: Emma's nickname for Charles was "Nigger." "My own dear Nigger," she would write in their correspondence—they wrote to each other frequently, even when living under the same roof. This nickname is probably a Victorian term of affection connected with slavery, denoting fond ownership. Still, a bit weird to our ears.

of medical training in Birmingham and London, but went on to pursue mathematics, and the world of numbers would be the prime determinant of his intellectual legacy.

Over his long and varied career, one thing was consistent among Galton's traits: He coveted data. He measured. It was in the statistics that he developed, and in his unquenchable thirst for measuring human characteristics, that he tried to formalize and lock down human differences. In Chapter 4 and elsewhere in this book, we explored the new business of genetic ancestry, where for around £100 and a froth of spit in a test tube, one of many companies will draw a sketch of your DNA. The results are, to my mind, of inconsequential interest to an individual, but in collecting these samples the companies behind them, notably 23andMe, are amassing colossal datasets of human genomes in numbers that far outstrip ones available to academic scientific research.

Galton had done it all before. He recognized the power of large collections of measurements—we call it "big data" nowadays—and cannily also recognized our own fascination with ourselves, and willingness to reach into our purses to satisfy those egos.

The huge spectacle of the International Health Exhibition in 1884, at the site now occupied by the Science Museum in South Kensington, featured a replica of an insanitary street, many of the new-fangled drainage systems designed to improve public health, and electrically illuminated fountains. Four million people came to see the new science on show.

Galton was there too. He set up an Anthropomorphic Laboratory, where customers would pay 3 pennies (almost a dollar in today's money) to enter, and anonymously fill out a card (with carbon copy for Galton to keep the record) with personal details. "The object of the laboratory," he wrote in *Journal of the Anthropological Institute* after the expo had ended, "was to show the public

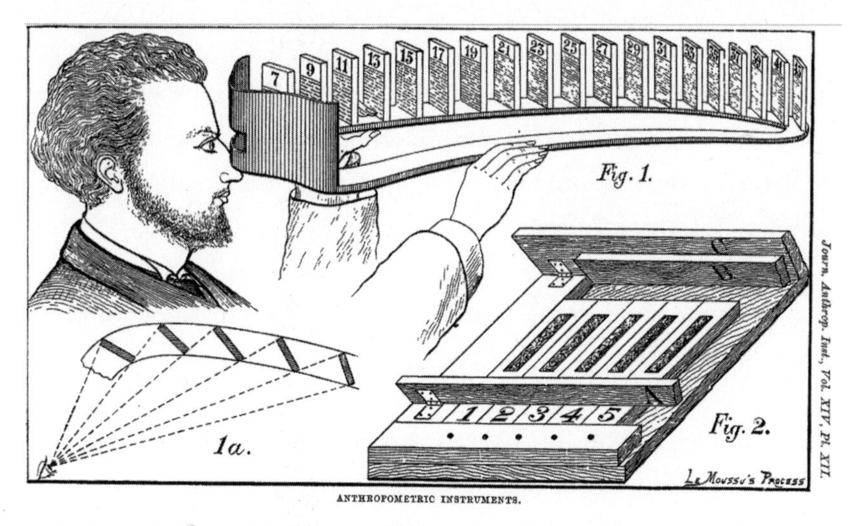

Galton's biometric equipment.
Figure 1 shows a sickle-shaped layout for testing acuity of vision, first with one eye and then the other. The pages are from the Shilling Diamond edition of the Prayer Book (though he points out that familiar texts are "objectionable" to the test, and—ever the mathematician—"a page of logarithms would be better"). Figure 2 depicts the "color sense" test, in which the participant is asked to identify shades of green wool in the five lengthways cartridges.

the simplicity of the instruments by which the chief physical characteristics of man may be measured and recorded."

They would file through, and be measured and tested for all manner of things. Some metrics were straightforward—height, hair color, arm span, and weight. Others were complex—keenness of sight, punch strength, color perception, and ability to hear high-pitch noises, tested via whistles made by Messrs. Tisley and Co., Brompton Road, and Mr. Hawkesley of Oxford Street. "Hardly any trouble occurred with the visitors." Galton noted, "though on some few occasions rough persons entered the laboratory who were apparently not altogether sober."

What a clever scheme. Today, people part with their cash to get some information about their DNA, just as they did for Galton at the Anthropomorphic Laboratory. Our love of navel-gazing means that we not only volunteer to give up our biometrics, but we actually pay for the privilege. Altogether, 9,337 gave up their thruppence to be measured by Galton, who was amassing data on people, hoarding the similarities and differences between people in all of these characteristics.

Mark Twain wrote in 1869 that "travel is fatal to prejudice, bigotry, and narrow-mindedness." Galton had explored extensively in the 1840s, as privileged young men often did in the nineteenth century, to Turkey and through the Middle East and Egypt. He went further, into what is now Namibia, on a two-year trip with the Royal Geographical Society, and published bestsellers describing his journeys into the heart of darkness. But Galton didn't adhere to Twain's maxim. He maintained and grew a deep-rooted sense of hierarchies of the peoples of the world, and formalized it later in his life under a number of auspices. A few years later, in 1873, he submitted a diatribe in the *Times* about how Africa should be best served if colonized by the Chinese as the characteristics of the "negro race" were not sufficient to foment development in their nations without help from the British. Recognizing that some "negroes" had accumulated wealth and demonstrated decent intellectual chops in a manner that, he asserts, is common among "Anglo Saxons,"

> average negroes possess too little intellect, self-reliance, and self-control to make it possible for them to sustain the burden of any respectable form of civilization without a large measure of external guidance and support.

Therefore, he suggests that as colonies of China, Africa would flourish, for

> The Chinaman is a being of another kind, who is endowed with a remarkable aptitude for a high material civilization. He is seen to the least advantage in his own country, where a temporary dark age still prevails, which has not sapped the genius of the race.

(though not before pointing out the bountiful negative characteristics of "the Chinaman," which include deceit, unoriginality, and timidity. "The Arab," for comparison, he concludes, "is little more than an eater up of other men's produce; he is a destroyer rather than a creator, and he is unprolific").

Though wince-inducing now, these sorts of views were not necessarily normal and not necessarily uncontroversial then, and we shouldn't assume that these were universal British Victorian values. Racism predated Empire, and these were the end times for slavery, which had peaked before Victorian times, even though the British Empire was still robust and proud. William Wilberforce had driven the Abolition of the Slave Trade Act through Parliament in 1807, which largely banned slavery, though only in 1833 was this extended throughout the whole Empire with the Slavery Abolition Act.

The Darwin-Galton-Wedgwood family tree is an impressive canopy. Upward from Charles is Josiah Wedgwood, founder of the pottery dynasty, and Erasmus Darwin. Alongside Charles was Galton, and the descent of Darwin features brilliant scientists, writers, actors and, with marriage into the Keynes family in the 1880s, economists. Galton was aware of his pedigree, and was moved to examine human excellence in families after the publication of *On the Origin of Species*. Darwin had boiled the bones of fancy pigeons, each displaying absurd and absurdly different plumes or gullets or feet, and in doing so had showed that they were the same species, and that these traits had been bred into the birds by pigeon fanciers for competition over thousands of years. It was crucial for the theory because it showed that species were malleable, not cast immutably in stone never to change. Galton wasn't thinking about feathers, but of ability and of genius—not least that which was clearly in abundance in his own clan.

Ten years after his cousin changed the world, Galton published *Hereditary Genius*, in which he carved out the idea that men of eminence—and it was predominantly men—ran in families:

> The general plan of my argument is to show that high reputation is a pretty accurate test of high ability; next to discuss the relationships of a large body of fairly eminent men—namely, the Judges of England from 1660 to 1868, the Statesmen of the time of George III, and the Premiers during the last 100 years—and to obtain from these a general survey of the laws of heredity in respect to genius. Then I shall examine, in order, the kindred of the most illustrious Commanders, men of Literature and of Science, Poets, Painters, and Musicians, of whom history speaks. I shall also discuss the kindred of a certain selection of Divines and of modern Scholars.

The measure of reputation is not part of the metrics that includes yards and feet, or volts and amps. So the starting point is questionable. To his credit as a good scientist he did recognize limitations in his methodology, and turned to twins as a means to study the relationship between biology and environment, nature and nurture. Darwin read *Hereditary Genius*, or more precisely, his wife Emma read it aloud to him, and ever polite, sent effusive letters offering praise on his clarity.

Galton suggests that the betterment of society could be drawn from understanding the role of inheritance of abilities as described in analysis of great men from history, and that the weak could be stored celibate in monasteries or nunneries. In 1883, he invented the word *eugenics*. It was from these works and from his experiences on his travels that Galton used to formulate ideas about why some people were better than others, why some were successful and others not, ideas articulated so clearly in that letter to the *Times*. By the last decade of the nineteenth century, politicians and thinkers were expressing concern that the British "stock" were not fit enough to fight in the Boer Wars, which had been raging in Africa since the 1880s. They turned to Galton.

His influence continued for decades. Winston Churchill attended a Galton lecture in 1912, and a few years later raised the topic of British stock in Parliament. Both he and Theodore Roosevelt desired the neutering of the "feeble-minded," as was the parlance in Edwardian days for all manner of psychological, cognitive, and mental health conditions. Roosevelt, not yet president, expressed the view that purifying the American human stock would be "for the benefit of civilization and in the interests of mankind."

Marie Stopes is known today as a champion of women's reproductive rights, and her name adorns hundreds of clinics worldwide that provide essential support for women and their choices regarding pregnancy. But she held some horrifying views, arguing forcefully for the compulsory "sterilisation of those unfit for parenthood," particularly the Irish in London. During the 1930s, she wrote love poetry to a rising European politician in praise of his policies, which included reform of his country's population structure using eugenics as part of their radical plans. His name was Adolf Hitler.

Support for eugenics spanned the political spectrum. William Beveridge, principal architect of the welfare state, whose ideas would form the foundations of the National Health Service, said:

> Those men who through general defects are unable to fill such a whole place in industry are to be recognized as

> unemployable. They must become the acknowledged dependents of the State . . . but with complete and permanent loss of all citizen rights—including not only the franchise but civil freedom and fatherhood . . .

George Bernard Shaw, also on the political left, said, "The only fundamental and possible socialism is the socialization of the selective breeding of man."

The British never did adopt a eugenics policy, despite England being the intellectual birthplace of the idea. Before Darwin and Galton, Thomas Malthus had formally fretted about population growth and control, and therein laid the foundations of improving the "stock" of a people. But in the USA, and a few other countries (notably Sweden), the forced, involuntary, and often secret sterilization of undesirables was embraced enthusiastically. From 1907, when Indiana passed the first mandate, until 1963, forced sterilization was legally administered in thirty-one states, with California the most vigorous adopter. The most recent cases of forced sterilization in that famously liberal state occurred in 2010. In the twentieth century, more than 60,000 men and women, though mostly women, were sterilized for a variety of undesirable traits—men frequently to curtail the propagation of criminal behaviors. Native American women were forcibly sterilized in their thousands, and as late as the 1970s, black women with multiple children were being sterilized under the threat of withheld welfare, or in some cases without their knowledge.

These horrors were all to come in the twentieth century, as were the fullest repercussions of population control via murder and sterilization that occurred during the Holocaust. The Nazis slaughtered not just Jews in their millions, but also homosexual men, Roma, Poles, and people with mental illness.

And to little effect from the point of view of eugenic purification. The Nazis murdered or sterilized more than a quarter of a million with schizophrenia in order to purge them from the German people. The incidence of schizophrenia was measurably lower for a few years after the fall of the Third Reich as a result, but by the 1970s it was unexpectedly high. Schizophrenia is a condition that is associated with dozens if not hundreds of genetic variations, and many people have many of them, without suffering any mental health issues. Furthermore, a high proportion of schizophrenics do not have children, and so a eugenic program appears to have had little long-term effect.

That word, *eugenics*, did not carry the toxic meaning it has today. At the beginning of the twentieth century, Galton's influence was formalized with the formation of the Eugenics Records Office in 1904, with him in charge. It was part of University College London, the world's first secular university, the first to admit women, the first to employ a Jewish professor,* and a bastion of progressive thinking free from the shackles of religious doctrine. After his death in 1911, it was renamed the Francis Galton Laboratory for National Eugenics.

By the time I studied there in the 1990s, it had long since dropped that noxious word to become the Galton Laboratory of the Department of Human Genetics & Biometry. For three decades it was housed in a lackluster 1960s building on Stephenson Way, just north of UCL's main campus, and typical of universities around

* Who, purely by coincidence, was Jacob Cohen, maternal grandfather of Rosalind Franklin, whose work was the foundation of Crick and Watson's landmark work in determining the structure of DNA. In another delicious coincidence, the last published work by Charles Darwin, in *Nature* in April 1882, concerned some work by a Nottingham-based amateur scientist on barnacles and freshwater winkles; his name was Walter Drawbridge Crick, grandfather of Francis. If biological family trees are messy and webbed, academic family trees are often bizarrely matted.

Britain, linoleum floors, frosted glass, and orange Formica panelling. The majority of my lectures were in the Galton Lecture Theatre, and that's where I first did my scientific research as an undergraduate. Every day, I mooched past a humble glass display case of some of Galton's biometric tools (to which I was largely oblivious), including a head measuring crank, a device for measuring nose straightness, and a copy of his female beauty map of Great Britain (the most attractive women according to his criteria were in South Kensington, the least in Aberdeen).*

The genetics department has now moved half a mile south, across the Euston Road to Gower Street, into the Darwin Building marked with a blue plaque; Charles lived there during the *intervallum* between the *Beagle* and his long-term residence at Down House, where he carved out his masterpiece. The building that housed the Galton Laboratory is now largely abandoned, apart from a few offices, one of which hosts the Galton Collection,† an assortment of his letters, notes, and scientific equipment.

It is interesting to draw these comparisons between two nineteenth-century great men of science. Darwin gave us the manner by which we evolved. Galton built the foundational tools for the study of inheritance. He introduced studies of identical twins as a means of extracting the biological from environmental, nature and nurture.

The two men are in many ways perfect opposites. Darwin was dogged by humility; Galton was arrogant. Darwin was beset by doubt; Galton was determined to prove his own ideas. Darwin invented evolutionary biology; Galton founded and formalized

* In his pocket, he carried a tool he called a "pricker"—a thimble with a mounted needle, so he could mark female beauty on a paper simply by fumbling in his trousers, and thus absolutely avoiding the appearance of being a massive pricker-wielding pervert.

† Which, purely by coincidence, my sister now helps curate.

many aspects of the biological study of humans. They both worked at a time when great leaps were being made in the study of life, which would lead to further unifying theories of biology. The great nineteenth-century Moravian scientist* Gregor Mendel's work from exactly the same midcentury time, though ignored until the beginning of the twentieth century, described the rules of inheritance—how characteristics pass down the generations from two parents to one child.

In the first few decades of the twentieth century, primarily at UCL, a new breed of biology emerged that combined statistics and Darwin, and formalized the mechanism by which evolution by natural selection occurs. Not long after, DNA was established as the bearer of the genetic material and, in 1953, Crick and Watson revealed that it was constructed like a twisted ladder, the double helix, which not only was built to be copied from generation to generation, but also harbored information, a code that could be replicated every time a cell divides. Many scientists over the 1960s cracked that code, and showed it to be a means of writing and recording the instructions to make proteins, and this gave us an understanding of why diseases occur, and why people look different from each other: A tweak in the genetic code results in a subtly different protein, which may have a visible effect—eye color, skin color, the curl of hair.

In that century of wonderful, world-changing, glorious science, we had biologized difference. In measuring physical characteristics and subsequently analyzing them in large numbers, Galton set the framework for what was to come: biometrics—the measurement of humans. With genetics, we could peel back the skin and reveal the root causes of difference, and place them in time, in geography, in

* Mendel was also a monk, as is often mentioned. I prefer to refer to his world-changing sciencing, rather than his competent monkery.

evolution. We armed ourselves with the tools to scrutinize the difference between individuals and groups of individuals with molecular precision. Surface characteristics—the visible phenotype—would be replaced with the data that underwrote those traits, altogether a more fundamental measure of the true differences between people—the genotype. Mutable and impermanent differences such as nationality or religion could be tucked away as crude anachronisms because DNA and genetics would furnish us with definitive answers to the questions of similarity and difference.

The genetics of race

Well, it did and it didn't.

Biology moved from the shape of bones and features—the morphological—to the molecular long before DNA became the star of the show, in fact well before DNA was even suggested as the material that transmits inheritance. Blood groups began to be established into the A, B, O system we still use today during the First World War. In 1919, the Polish scientists Ludwik and Hanka Hirschfeld looked for a pattern of blood group distribution among soldiers from sixteen different populations they examined (mostly by country, but Jews were also included as a group). What was being measured and differentiated there were proteins, a proxy for the DNA that encodes them, expressed as alleles of the same gene, and they found variation in the frequencies of the different blood groups across their samples.* The same system was employed in the

* Ludwik Hirschfeld revealed his own prejudices and stereotyping in the methods he used to coax the blood samples from soldiers. "It was enough to tell the English the objectives were scientific," he wrote in his autobiography. "We permitted ourselves to kid our French friends by telling them we could find out with whom they could sin with impunity. We told the Negroes that the blood tests would show who deserved leave, immediately, they willingly stretched out their black hands to us."

1970s by the geneticist Richard Lewontin who examined blood groups with a far greater degree of precision, enabled by the emerging field of molecular biology. He looked at a hundred alleles across the genes that constitute blood groups—the subtle differences between people in the same gene. In his landmark 1972 paper, Lewontin quantified the precise differences between the molecules of blood type between peoples, and demonstrated that the highest proportion of genetic differences were seen within racial groups, not between them. Eighty-five percent of human variation, according to the genetic differences in blood groups, was seen in the same racial groups. Of the remaining 15 percent, only 8 percent accounted for differences between one racial group and another.

These numbers have been replicated in other studies of other genes since. What this means is that biology fundamentally deceives our eyes. Genetically, two black people are more likely to be more different to each other than a black person and a white person. In other words, while the physical differences are clearly visible between a white and a black person, the total amount of difference is much smaller than between two black people. If everyone on Earth was wiped out except for one of the traditional racial groups, say, eastern Asians, we would still preserve 85 percent of the genetic variation that humankind bears. They might look more homogeneous, but that fact reveals that the underlying code that causes the characteristics that we use to broadly define race has a disproportionately visible effect. These morphological differences are real, we all know that, but they're not representative of the genome as a whole.

Of course, there are differences: visible, measurable, and cryptic. It would be intellectually dishonest not to acknowledge them. People from eastern Asia have a darker skin tone than Europeans, as well as having thicker, black hair, and the epicanthic fold, which gives them an eye shape unlike others, and is largely absent in the people of the rest of the world.* In genetics we look for the underlying causes for these variations. Some are merely genetic drift—coded DNA that has changed over time and become fixed in a population not because they are useful, but simply because they were present in the individuals from whom the present-day populations are heavily drawn. Other characteristics may well have a selective advantage, but they are notoriously difficult to prove. Pale skin, prompted by just a couple of genes, is almost certainly an adaptation to lower levels of sunlight, and the subsequent reduced ability to generate vitamin D in the relative gloom of Europe (all discussed in Chapter 2), as well as a host of other proposed or theoretical factors, such as higher resistance to frostbite and simple preference in mating. There is natural variation in whom we fancy, and this might not be solely determined by some form of unconscious selection of a trait that will enhance success in one's life (such as food foraging skills, or the muscles to fight off a sabre-tooth cat). It might well be that you simply have a slight preference for people with a particular trait—ginger hair, for example—and as a result of that the genetics that code for ginger hair will persist.

The media (and to a lesser extent scientists) tirelessly speculate about the evolution of particular traits or behaviors in humans, with neat explanations as to what the advantage they provided was. Many are silly, and ridiculously unscientific—women like pink because as the gatherer half of hunter-gatherer cultures, being able

* Largely absent, though present in many other populations of non–East Asian descent, including Berbers, Inuits, Scandinavians, Poles, and indigenous Americans. Furthermore, there are visible differences in epicanthic fold shape between the different people of Southeast Asia to the extent that many people can distinguish between Koreans, Chinese, Japanese, and so on, solely on this basis. People with Down syndrome frequently exhibit epicanthic folds that were thought in the nineteenth century to be similar to those of the people of Mongolia. Hence, those with Down syndrome were known as Mongoloid or Mongol for more than a century. "Mong" was a playground insult when I was growing up, but is now considered offensive, and no sensible people still use it.

to spot berries would be useful; babies cry at night to prevent their parents from having sex and therefore create competition for them in the form of siblings.* We label these pseudoscientific fantasies adaptationism, or sometimes a form of panglossianism, after Dr. Pangloss from Voltaire's *Candide.* An eternal optimist, he suggested there was a reason for everything, and everything had a reason. Hence our noses were shaped as they are to hold glasses in their place, and we had two legs because that perfectly suits the structure of a decent tailored trouser.

The unglamorous truth is that there are but a handful of uniquely human traits that we have clearly demonstrated are adaptations evolved to thrive in specific geographical regions. Skin color is one. The ability to digest milk is another, which fits perfectly with the emergence of dairy farming (again, discussed in Chapter 2).

The best-understood example of regional adaptation concerns the single greatest cause of death in the human story. Genetic mutations emerged in populations in malaria regions that offer some protection against two versions of the disease, caused by the single-celled parasites *Plasmodium falciparum* and *Plasmodium vivax.* An alteration in the gene hemoglobin-B causes a structural change to the shape of blood cells, normally a half-sucked lozenge, which becomes rigid and curved like the blade of a sickle. People with one copy of the mutated gene have sickle cell trait: Their blood contains some of the distorted blood cells, but they are largely unaffected. People with two copies have sickle cell anemia,

* These are both real suggestions from academic papers. The pink thing is complete nonsense for a couple of reasons: the first is that the gender preferences for pink are a twentieth-century phenomenon. The color schemes in Victorian children's bedrooms were frequently pink for boys, blue for girls. The second is that there is no evidence that women acted as the gatherers and not the hunters. As for babies crying to prevent parental sex, yes, it does have that effect, as every parent will know. I'm not sure it's the definitive reason though.

a serious disease that 300,000 children are born with every year, with symptoms including pain, infections, and increased risk of stroke, and death, all because the misshapen blood cells can get lodged in blood vessels and organs. But sickle cell trait offers protection against infection from malaria. And so, when we look at the distribution of the mutated gene, it perfectly matches the range of malaria all over the world. Furthermore, some researchers have suggested a strong association with the presence of the protective gene in populations who have historically farmed yams. To plant yams, farmers clear forests. Cleared forest means more standing water. More standing water means more mosquitoes. More mosquitoes means more malaria—so the idea goes. The emergence of the disease, and as a consequence the resistance gene, may well have been enabled, or at least nurtured, by yam farming. The persistence of sickle cell anemia is the cost of positive selection for resistance against the most destructive disease in our history.

Interestingly, sickle cell is culturally often thought of as a "black disease," even to the extent that it comes up in the lyrics of hip-hop frequently, sometimes as a diss.* Sickle cell trait and sickle cell disease are not exclusive to black people, not least because the term *black* is utterly useless as a descriptor of people whose origin is in Africa. We've already established that the vast majority of genetic diversity occurs within that continent, as this was the pool from which only a small sample of humans wandered out to populate the rest of the world. While it is true that the densest concentration of the sickle cell gene is in Sub-Saharan Africa, it is prevalent in the people of every other region that is affected seriously by malaria, including the Middle East, the Philippines, South America, and southern Europe, notably Greece. It seems that disease and evolution have little regard for continents or racial epithets. For the most

* For example, in the impressively angry 1992 B-side "Hit 'Em Up" by Tupac.

part, it is not known how adaptive the characteristics we see in specific regions are. That doesn't mean they haven't been selected.

Hitchhiking, surfing, and sweeping

Earwax is of great interest to people like me. Stick your little finger in your ear and have a good root around. The stuff that sticks to your finger is called cerumen, and it's a mixture of dead cells, fluff, dust, and other detritus of a life. We like it because it's one of a very small handful of traits that has a relatively straightforward relationship between the DNA and its outcome—the genotype and the phenotype.

There are basically two types of earwax, sticky and dry. The gene that determines these two states is called ABCC11, which comes in two alleles to give them their more scientific and less revolting descriptor. The gene is 4,576 base pairs long, and at position 538 there is either a G or an A. If you have a G, the code writes the amino acid glycine, and if you have an A you get an arginine. This simple change slightly nudges the protein into a different shape, and the shape switches the nature of the wax. The inheritance of this terribly important phenotype plays out in a human ear in a straightforward Mendelian way. Wet is dominant: two copies of the G version and you have wet earwax; one of each allele and you have wet earwax; two copies of the A version and you have flaky, crumbly dry earwax. This charming business gets much more interesting when we look at how earwax is smeared across the world.

We think mostly not about the phenotype, but the genotype in genetics, and measure the frequencies that the different alleles have in populations. In the case of earwax, the two relate to each other very directly, so, for example, if the dry allele occurred at a rate of 50 percent, in a population of one hundred people, you'd expect to see twenty-five people with dry and seventy-five with wet wax (not 50:50, because those with one wet and one dry gene have wet wax). In Africa, the proportion of dry genes is effectively zero. In South Korea it's the other way round. Generally, eastern Asians have a far greater frequency of dry than anywhere else on Earth, and broadly the further east you go, the more likely you are to find dry wax ears, if indeed inspecting people's ears is your business.

Now, why on earth would this be? Drift might be one explanation—that founding populations of the east had more dry genes than the ones that they left behind in the west as they migrated, and that for no particular reason those proportions spread and became fixed. We can look more closely with genetics, and here is how we do it. Genes occupy locations on chromosomes, but geneticists don't tend to look at whole genes. They look for SNPs, those individual letters within genes that vary between individual people. These aren't typos, as they don't cause disease, they're just variant spellings—*skeptic* or *sceptic*, *grey* or *gray*. When tracking the evolution of people, you don't just look for one SNP, you scan the DNA around the bit you're interested in for others, because DNA is passed from generation to generation in chunks. A gene that is of advantage to the individual may be selected with flanking bits of DNA in tow.

As well as being a playwright and a vocal eugenics advocate, George Bernard Shaw was a linguaphile, and described the USA and the UK as two countries separated by a common language.* Here are five words with two alternate spellings, some different in UK and US English (but the same number of letters): *grey*, *disk*, *barbeque*, *theatre*, and *adviser*. Here they are in a nonsensical sentence:

* Probably. The earliest incarnation of this thought is in Oscar Wilde's 1887 jolly horror story "The Canterville Ghost": "Indeed, in many respects, she was quite English, and was an excellent example of the fact that we have really everything in common with America nowadays, except, of course, language."

> Your grey disc is a theatre barbecue adviser.

Here's an alternate version, equally valid (and equally nonsensical), but with five legitimate alterations:

> Your gray disk is a theater barbeque advisor.

If each of the variants were independent of each other, and you sampled everyone in the world's spelling, then you'd expect to see every possible combination of those five variants in the sentence, twenty-five in total, at the same frequency. If some of the words in the sentence are linked to another, for example the adjective *grey* and noun *disc*, then you might see these two together more often than not. In genetics, this is one of the key ideas, with a typically wretched and obfuscating technical name: linkage disequilibrium. *Grey* and *disc* are linked because they associate with each other in location and meaning, an adjective relating to a noun. In DNA the association has no inherent meaning, but the proximity between two SNPs is all important. We look for variations in DNA that associate with each other in clumps. With the easy access of the full human genome these days, we look not for five differences, but tens, hundreds, thousands, hundreds of thousands. Specific SNPs can undergo linkage disequilibrium as a result of selection (that is, both variants are useful), or by what's called "genetic hitchhiking"—one variant is useful, and others that are sitting nearby get dragged along into future generations. Because *grey disc* is preferable to my British eyes, *theatre* is coming along for the ride too. They have been linked and are now not inherited with an equal probability. They are more likely to be inherited together. These statistical blips are what tells us how we have evolved.

Another related effect that can occur is called a "selective sweep." This is when one particular variant confers an advantage for the organism, and as a result of hitchhiking over many generations, all other variants are removed until all the variants are deselected, and there's only one version of the full set: *Your grey disc is a theatre barbecue adviser*—they have all been linked now.

These concepts are key tools in the geneticist's toolbox. By taking DNA samples from people spread out over the world, we can use the differences we see in linkage disequilibrium and selective sweeping to suggest patterns of human migration. The patterns are subtle, and it requires delicate math to get these ghosts to emerge. They are hidden in living people, so the (perfectly reasonable) assumption is that the geographical spread shows a degree of recent permanence. It's not unreasonable because for most genetic variations examined so far, we see them change gradually over landmasses, and more abruptly at seaboards. Although modern movement of people is rapidly messing up these genetic shadings, we can still extract country of origin for previous generations.

Back to earwax. A study in 2011 suggested that with these techniques, the earwax gene had undergone a positive selection as it has moved east. That isn't necessarily to say that it bestows a clear advantage to those who have it—it's not easy to even speculate in a panglossian way what benefit flaky ears might bestow. It may merely be positive selection as a migrating population spreads into a new area, has more kids, and the new gene frequency is different from the place whence they came. This is an example of riding the wave of migration, what some geneticists call "surfing." It might be that it has hitchhiked its way into the east along with something unknown but of much greater benefit.

Japanese researchers in 2009 suggested a reason why the global distribution of earwax is spread the way we see. They proposed it is because we smell. Your axilla is the hairy bit where the arm joins the chest, and we spend billions each year masking the smell that this area naturally generates with deodorant. The wet wax type occasionally associates with a disorder called axillary osmidrosis in a way that is much more rare in our flaky-eared cousins. People with this condition believe their armpits stink, and sometimes seek surgery to remove the glands that generate the sweat that is behind the odor. It seems that ABCC11 may have a role in these apocrine glands, though we don't know what. Could it be that what was actually being selected was smell (or lack thereof), and the dry wax was merely hitching an evolutionary ride? I am possibly committing what I decry above: This is adaptationist speculation. It may be true, but we simply don't know. So for now, this remains merely an interesting just-so story. It might render the people of South Korea the least smelly on Earth, though this clearly important study has not yet been published.

While we're in the Far East, there's another example of so-called racial characteristics, more visible than earwax, but also concerning sweat, which also have become fixed the further east we travel from Greenwich, and that have also surfed their way into being putatively racial. This one is a gene called EDAR. It's of similar but different use to us here because it doesn't just do one thing. The gene EDAR sits on the second human chromosome, and encodes a protein called ectodysplasin A receptor, which typically lies on the surface of some cells of the developing embryo. There it plays a role in communicating between two major classes of cells in the growing flesh of a body, the ectoderm and mesoderm. In this interplay, many tissue types are defined, including hair, teeth, nails, and sweat glands. As with so many human genes, we know how it works by analyzing what happens when it's broken, and in this case, mutant versions of EDAR cause disorders such as hypohidrotic ectodermal dysplasia, where patients have few or no sweat glands, no hair or fingernails, and peculiar teeth.*

* Horror movie fans will well know the face of Michael Berryman, whose

Of the normal versions—the alleles—of EDAR we see in humans, one, referred to as 370A, is almost omnipresent in East Asians and Native Americans, and almost absent in Europeans and Africans. This particular allele associates with thicker hair, an increase in the density of sweat glands, and a particular front teeth shape called "shoveling" (it's not visible from the smile, but the back of the incisors is scalloped away in a particular way). "Associates with" is not the same as saying "causes," because it is very tricky to show precisely that a genetic mutation encodes a protein that is altered in such a way that it produces these changes during growth in the womb. Nevertheless, experiments in mice indicate that these characteristics are the result of the 370A mutation.

In 2013, a neat international study did a range of things that set the bar for how to quiz the genetics of our evolution. Geneticists from MIT, Harvard, Fudan University in China, and of course, Galton's own alma mater, UCL, scanned the genomes of 1,064 people from fifty-two populations around the world, and looked for SNPs that flanked the 370A allele. They ended up with a block of DNA consisting of 139,000 bases, with 280 other SNPs of interest. That's the ease with which we can now dip into the genome. It has tons of genetic data, and crucially of genetic variation, to play around with to question its origins in human history, and what it enables is the ability to calculate when and where 370A originated. Recall that this is not a common allele in Africa and Europe, so was acquired by an entirely random process at some point as people moved east. The computer simulations include a whole swag bag of

---

striking features made him a staple freak in films such as *The Crow* and *Weird Science*, and a mental health inpatient in *One Flew over the Cuckoo's Nest* in 1975. That look, leering down from the VHS cover of the 1980s horror film *The Hills Have Eyes* (and its lesser sequel) is classic hypohidrotic ectodermal dysplasia, and will be terribly familiar to anyone who, like me, haunted video rental shops in their misspent youth.

input variables, including farming practices, migration, the euphemistic gene flow between distinct populations and other factors, and what comes out the other side is a number. Or, rather, because science is ultra-cautious when it comes to statistical analysis, a range of numbers. What emerges is that this mutation occurred in an individual somewhere between 13,000 and 40,000 years ago, but the most likely date is 31,000 years before today, in what is now central China. This predates the migration across the Bering Strait (which was then terra firma) by the people who would populate the Americas north and south.

The second part tested what the mutation actually does, and they used the standard of genetics for several decades, our old friend the mouse. When a gene is faulty in humans, once identified, it gets tested in mice by inducing the same mutation, and seeing if the result is similar or the same. When the gene is rendered functionless, this is called a knockout, because you eliminate the gene you're interested in. The opposite test is also useful, a knock-in. A version of the mouse's EDAR gene was knocked-in with the 370A mutation in place, and it had thicker hair, denser sweat glands, and smaller, more branching mammary glands. Rodents' incisors grow continuously as they gnaw them down on food, wiring, and skirting boards throughout their lives, so the shape of the teeth is not an informative comparison between us and them. But the other phenotypes were remarkably similar to what we see in East Asian humans.

Speculation as to the reasons why this mutation so successfully spread was rife. Sweat gland density may well relate to climate, as these glands form an effective cooling system in hot and humid conditions, especially for long-distance running or walking, and this might've been desirable for the culture of hunter-gathering people of this period. Geological records show China was indeed hot and humid at around the time the mutation arose, and may have retained this heat due to monsoons at a time when temperatures were broadly dropping. One way to address this would be to do more digging. Discover bodies from the time and get their DNA out, and dig up more of China to get better historical climate records. In science, there's always more digging.

Or it could relate to smell, as suggested with the earwax gene. As ever, speculation in the press looked to the most exciting possibility, and some articles suggested that reduced breast size was visible and therefore possibly a driver of sexual selection that helped fix the new mutation into place. This is pretty flaky science. Breast size is not universal in being a determinant of sexual attractiveness, despite what some strands of contemporary western culture insist upon, and neither mammary gland size nor the internal structure of the ducts of lactation are direct correlates of bra size. But why let the facts get in the way of a good story, especially when it's got a sexy headline.

Whatever the selective advantage was remains unknown. It is unlikely that all characteristics were advantageous at the same time, and of course we have to try hard not to think of this evolution in a clear linear way. It may be surfing, hitchhiking, or sweeping, and the length of time it took to become omnipresent will be thousands of years and thousands of generations of people. Dan Lieberman, a Harvard professor of evolutionary biology who helped conduct the study, told the press at the time that, "These findings point to what mutations, when, where, and how. We still want to know why."

What it shows more clearly than ever before is that key physical attributes that we identify as being "race-specific" are superficial and recent. Even with the relatively straightforward EDAR gene, we still don't really understand why it became so prevalent among more than a billion people, and stayed that way. You could breed out the 370A from a family in a few generations simply by a

Taiwanese man marrying a European woman, and their children marrying Europeans until the only allele in that family is the original version. It could conceivably be lost in two generations, with both copies eradicated in successive generations. The children in that family would not have thick black hair, and would have acquired sweat glands like Europeans or Africans. What race are they then? Equally, you could reintroduce it in two generations by marrying into a Vietnamese family. Are they back to being East Asian then? The genetics is not definitional and not essential, and refuses to align with the way we talk about race. And this is just one gene. Behavior is breathtakingly complex to scrutinize.

Yet a book published in 2013 by the former science editor of *The New York Times*, Nicholas Wade, made a number of questionable assertions relating to the genetics of race. *A Troublesome Inheritance* posed some ideas that race is not only very clearly defined genetically, but that these distinctions in DNA account for not just the physical characteristics of certain populations, but also some of the social and cultural behaviors.

The book frequently misrepresents much of the work that is used to defend his assertion that recent evolution within so-called races explains why certain people appear to be better or worse at certain things. According to Wade, the English display a "willingness to save and delay gratification" and this is absent from certain tribal cultures. Jewish genes are "adapted for success in capitalism." The Chinese are predisposed to obey authority (how similar this sentiment is to that of Galton expressed in the letter to the *Times* in the nineteenth century). These statements are unsupportable in any form based on our knowledge of history, genetics, and cognitive ability. They are also clumsy and gross stereotypes and, in my opinion, straightforward racism. But *A Troublesome Inheritance* received much press coverage because it laid out controversial and provocative ideas, purportedly based on scientific evidence.

Deconstructions in the form of demolition of the book have been universal among geneticists. But some of the fallacies and inconsistencies feebly presented as science are useful for understanding why this is such a sticky area to deal with.

There is no doubt that humans have evolved in the very recent past and are still evolving today. The significance of that change in genetic material over decades, centuries, and millennia is the subject of scientific debate. Wade frequently makes assertions about the changes in genes in his races over very recent human evolution that are simply unsupportable and, moreover, he does not try to support them. Let us not forget that evolution simply means change over time, so the question is really not if we are evolving but we are evolving under the duress of natural selection. Are we adapting to local conditions according to our genetic material? And if we are, do these adaptations correspond with the way we commonly describe races?

What is race?

What do we mean by race anyway? This is not an easy question to answer: Everyone thinks they know what race means, and can tell the difference between peoples with varying degrees of granularity. The epicanthic folds typical of East Asians are variable within East Asia, and with experience you might be able to, on average, correctly identify a South Korean from a Cambodian. But nobody would classify these as different races. They're both East Asian. Inuits have an epicanthic fold too, but they're not East Asian, though we know they came from a population in East Asia. Black, as discussed above, is virtually meaningless as a scientific descriptor, and Africa as a racial group is also of very limited use because black people are more likely to be more genetically different from each other than they are from white people. Nevertheless, you might, with experience, be able to distinguish an Ethiopian from a Senegalese. That distinction, the shape of a face, the color of skin, the breadth of a nose, is precisely why "black" or "race" is a term of little value to science. We use black casually to mean "someone with dark skin whose recent origins have been in Africa" (as opposed to "someone with a similar skin tone from India"). There is not much precision in that statement.

Galton was quite comfortable with broad definitions and crude associated assertions: Hindoos, Negroes, the Arab, the Chinaman, and so on. Already we can see how imprecise the language employed here is. China is a country, so in a sense, the easiest to define—people born inside China. Hindu is broadly a religion or culture, though also generally refers to the largest proportion of members of the Indian subcontinent, and doesn't include the 180 million Muslim Pakistanis, nor Indian Muslims, both of whom are genetically indistinguishable from Hindu Indians. The Arab? For one so formal in measurement and in categorization, Galton shows his racist hand in being so relaxed about these definitions.

Early attempts at human categorization lumped us all into five, and this endured with remarkable stickiness. The eighteenth-century German anthropologist Johann Blumenbach described Caucasian, Mongolian, Ethiopian (broadly meaning sub-Saharan African), Malayan (roughly Southeast Asian and Pacific Islanders), and Native American. A simplified version arrived at the end of nineteenth century—Caucasoid, Negroid, and Mongoloid. The US anthropologist Carleton Coon stuck with five, but slightly different groups in the mid-twentieth century—Caucasoid; Mongoloid (which included everyone indigenous to the Americas as well as East Asia); Australoid (meaning aboriginal Australians); and Negroid was divided into Capoid and Congoid (from the Cape and the Congo, as a means of distinguishing sub-Saharan Africans of the east and south from central and west).

The minimum number of groups humans can be classified into has never remained static. Nicholas Wade himself vacillates between there being three and seven races on Earth today in his book, and equates them very broadly to continental populations—African, eastern Asian, and Caucasians, but might include Indian and Middle Eastern, and others. This is very broadly what many mean when they talk about race, though clearly it has underlying problems. We know that the emergence of the pale skin we associate with Europe, and particularly northern Europe, only emerged in the last few thousand years, just as the genes for processing milk did. We've addressed the single gene EDAR, and how it confers that characteristic thick black hair that is typical of eastern Asians. In terms of genetics, we're looking at a handful of genes among thousands, and minuscule factors of variation across millions in the whole genome. There is no single gene that underpins the concept of race, just like there are so few genes for any one complex human characteristic, and there are just a few that convey the broad physical differences that render populations very visibly different from each other. Even when there are, such as with EDAR, these represent a superficial and tiny fraction of the total amount of genetic difference between people.

The Human Genome Project moved the field on in a seismic shift, as we could scan not one gene or a handful, or one variation in spelling such as those first assessments using blood group proteins. Instead, we can scan for hundreds and thousands, in thousands of people. Noah Rosenberg from Stanford in California led one of the first major studies that did just that in 2002, and used the new power of genomics and mighty computing to scrape not just beneath the surface, but into the depths of our evolution.

His team sampled SNPs from 1,056 people, from 52 geographic regions, and looked at variations at 377 locations spread across the whole 3 billion letters of each of their genomes. This was a huge study at the time, and though 377 dots in an ocean of 3 billion sounds like a drop, that is plenty enough to determine the spread across nations. They fed this data into a computer program called STRUCTURE that sorts for similarities by clusters—plug the numbers in and then ask it to sort them into a number of categories that you decide. They asked it to give out a range of clusters from two to six—that is, divided into two groups of people based on similarities, then three, four, and so on. When they did two, it grouped all people into those from Africa, Europe, and western Asia, and those from eastern Asia, the Americas, and Australia. That makes sense; these geographically are the places that we populated first. When asking the algorithm for three groups, Africa is hived off as a separate group. That also makes sense; the people who went out of Africa were only a small sample of the total available alleles. With five groupings, Australia becomes a separate group, as does eastern Asia. All of a sudden, the genetics appears to confirm the most traditional racial grouping: African, Europeans and the Middle East, East Asians, Australians, and the Americas.

But push it to six and something weird happens. The Kalasha, a northern Pakistani tribe of around 4,000 people, emerge as the next group. They are a strange population, small and isolated in the highest mountains of the Hindu Kush where it's barren and glacial, and accessible only by mountain passes and rope bridges. They are largely endogamous—marrying within their community—and have their own language and religion, though many are now converting to Islam, as their neighbors across the border in Afghanistan, the Nuristani, did at the end of the nineteenth century.

The Kalasha are indeed an interesting people, and unusual in many ways. But they're not that unusual, and certainly not nearly unusual enough to warrant placing them in a whole separate racial grouping, as no one possibly would consider doing, not even the most ardent racialist thinker. If you keep increasing the number of clusters, you get more and more groupings of people, geographically and culturally bound. In fact, the harder you look for this fine-scale architecture within the 3 billion letters of genetic code that each of us has, the more gradients emerge at the boundaries. With these cluster analyses, it is true that the most similarities examined are shared within a group, but plenty overlap with other clusters. The graphical representations of these types of data show blending at the edges. The sharpest delineations coincide with water: Europe, sub-Saharan Africa, and East Asia. But with the addition of more groups, with fewer oceanic gulfs, human variation is pretty continuous. The concept of a discrete or pure race vanishes in the haze.

Rosenberg's study is a great piece of work and posited the right questions to interrogate the underlying genetics of how the people of the world are distributed. By scanning thousands of single positions spread across the whole genome of more than a thousand people, it reinforced—in fact amplified—Lewontin's earlier results that the biggest genetic differences in people were seen between people of the same so-called race, rather than between the races. But at exactly the same time, it showed that the differences we do see when picking through the genome with a fine-toothed comb are the same as the basic racial structure that had been suggested for decades, in the era before science, during Galton's time, during the years of crude surface traits, and throughout the twentieth century. This is not looking for the genetic differences that cause the visible differences between people, but simply hunting for any differences that might exist between them, visible on the outside or not.

Look for clusters, and you'll find clusters. The fine-tuning of where these clusters exist is fascinating and the question becomes "why?" Why do we see these groupings, especially if the rest of the genome, in fact the majority of the genome, does not show such regional variation?

We often rely on language as a metaphor for explaining genetics, and I'm going to attempt to use it here too. Imagine all the books currently in print in the world. To simplify, let us just refer to those written in English, and nonfiction. Publishers and bookshops like to categorize them in order to help promotion, to push sales, and to help the reader get an idea of what it is that they are buying. You're holding a science book, though it has plenty of history in it, and it's primarily a biology book. My last book,* which was about the origin of life, was also science with plenty of history, but while it had plenty of biology, it also featured physics, astrophysics, geology, and chemistry, as befits the study of the transition from inanimate chemicals to living systems on the young earth.

In the classics of science writing, Darwin wrote about geology, Carl Sagan was a cosmologist who wrote eloquently on biology and physics in his masterpiece *Cosmos*, as does the particle physicist Brian Cox in the modern era, and anatomist Alice Roberts frequently writes about archaeology and history. So while we can agree that this minuscule sample comprises nonfiction books in English, further classification is murky. If we then include all genres of nonfiction—from hokum fad diet books to the *Highway Code*, via car manuals and celebrity biographies—you begin to see the problem. But we're not meant to judge books by their covers, or titles, but by their words. Would this help the classification system? If we were to go through the text of all these books and look for the word *science*, it would appear most frequently in books about science, but not exclusively. Science (misused) is also part of the toolshed of fad diets and flaky spirituality guides. Context is essential.

So we expand the search criteria and use *science* and *biology*, which ought to refine the categorization somewhat. But there will be books that contain the word *biology* but not *science*. Do we include

* Which is called *Creation: The Origin of Life & The Future of Life* (Viking, 2013).

them in our *science* and *biology* grouping? Say we want to find books about biological evolution, so we might include *science*, *biology*, and *evolution*. Alas, Darwin himself does not use this word in the original text of *On the Origin of Species*, so that's not going to work. He does talk about barnacles a lot though, and the ship on which he traveled the world, the *Beagle*. And indeed there are multiple books—for example *Mystics Seafarer's Trail* by Lisa Saunders* and *Plant and Animal Alphabet Coloring Book* (1979)—that speak of the dogs and the shellfish, and not about Darwin, the explorer ship, or evolution. But these are not in the same category.

And so it goes. Booksellers do their best, and broadly we can say that there are science books, but on close scrutiny it's a fuzzy definition at best. Certainly the books that sit on the shelf in a bookshop alongside this one you are holding are more likely to contain the words *science*, *biology*, and *evolution* and so on, than the ones in the cookbook section, and so could broadly be lumped together into an appropriate category. But some of them are going to be physics books, or math books. And you wouldn't have to have gone far to the left or right on the shelf to encounter a "science" book crammed with pernicious unscientific flapdoodle. I do not know where the lines are drawn.

We do this all the time with all things. In art, there's cubism and Dadaism and surrealism and public art and video installations and portraiture and photography. Politically we're left wing or right wing or conservative or Conservative or liberal or libertarian. Films are westerns or science fiction or horror or rom coms. And woe betide my ex-girlfriend who said she didn't like black and white films. It's not that there aren't measurable, quantifiable differences

* This was found, via Twitter, via the Google Books text search algorithm. I haven't read it. *Ulysses* by James Joyce has the character Molly Bloom, who is drawn from his wife, Nora Barnacle. *Ulysses* also mentions beagles, but of course that is fiction, and far more difficult to understand than genomics.

between all these categories we impose upon things, it's just that for the most part they fit not into discrete units, but into a continuum. We are naturally plagued by the tyranny of a discontinuous mind, as Richard Dawkins so eloquently said.

The analogy works up to a point. It fails to recognize that certain genetic groupings do roughly correspond to geography. But not exclusively, and not essentially. The analogy does though satisfy the question of how many races there are: It is unanswerable. It is a meaningless question.

Yet it never goes away. The idea that Native Americans were genetically predisposed to alcoholism persists today, as it has done since the early days of the European occupation of the Americas. Thomas Jefferson wrote a letter in 1802 to an Iroquois chief praising the tribe's decision to adopt abstinence from "spirituous liquors," and going on to say "as you find that your people cannot refrain from an ill use of them, I greatly applaud your resolution not to use them at all." Addiction is a staggeringly complex problem to understand as there are so many biological, social, and cultural factors to consider, including poverty, education, family history, and trauma experienced during childhood. And yes, genetics appears to play a role in about half the total risk for being alcohol addicted. But there's no evidence that Native Americans have any versions of genes that metabolize alcohol any differently from white people in America, nor is there a simple single genetic factor that might render someone an alcoholic. There is plenty of evidence for brutal social and cultural experiences for many Native Americans, and generations of oppression, resulting in underemployment, poverty, and low socioeconomic status, all of which are risk factors for alcoholism. Yet, the notion that the high rates of alcoholism in Native Americans—almost twice as high as in white European immigrant Americans—are somehow genetic remains an oft-repeated idea.

❧

In the 1880s, two doctors independently identified a new horrible disease, both of them finding it in Jewish families. In London, Waren Tay had spotted red dots in the retina of young children from a single Jewish family, and followed the disease progression through gradual nerve deterioration and death. Bernard Sachs also saw similar symptoms in New York, and proposed a name: amaurotic familial idiocy. The disease they were both seeing is now called Tay-Sachs and is well understood as a recessive disease caused by mutations in the HEXA gene. It's a terrible syndrome in which the brain deteriorates over a short time in very young children, and they die soon. Within a few years of doctors Tay and Sachs characterizing their disease, children with identical symptoms had been described in Gentile families, but already, because Tay-Sachs was a "Jewish disease," they were adjudged to have something different.

Tay-Sachs is not a Jewish disease. It's seen at roughly the same frequency in Cajuns in Louisiana, and French Canadians in Quebec. There's no such thing as a Jewish disease, because Jews are not a genetically distinct group of people. Certainly there will be higher levels of genetic similarity in families and in related groups, and indeed Tay-Sachs did for a long time have a higher frequency in Ashkenazi Jews than in some other groupings of people. But it's not exclusive to Jews or Ashkenazi Jews or Sephardic Jews or Cajuns or any single identifiable group of people. Yet the myth persists. When talking about race and genetics in public, as I sometimes do, frequently the question will be asked, "What about Jewish diseases, such as Tay-Sachs?"

Here's an irony: There has been a great deal of research into the genetics of Jewishness, much more than for many social groups. This is probably to do with a high proportion of Jewish geneticists and scientists in general, and the very unusual history of the Jewish people, whose diasporas and persecution have made them an interesting case study in the way genes and culture interact. Because of this interest, the prevalence of Tay-Sachs has been tackled. In Ashkenazi populations, which make up around a third of Jews in total, careful genetic counseling has effectively eradicated Tay-Sachs. I suppose you could call this a form of soft eugenics in its purest nonjudgmental sense. It was called a Jewish disease at first and that stuck, carried along by prejudice and ignorance. Now, because of an understanding of genetics and inheritance, it most certainly is not a Jewish disease at all.

In sport, similar ideas doggedly persist, despite science. There hasn't been a white man in the Olympic 100 meters final since Allan Wells won in Moscow in 1980. African American athletes have provided thirteen of the top twenty speeds in the 100 meters in history (the other seven were also by black men, Canadian or Jamaican), and they boycotted the games that year, as the Cold War was as chilly as it would get. These types of numbers have fuelled a notion that the prowess and success of black people in sports is a biological, and therefore genetic advantage that they have over white athletes. Recall Jesse Owens standing on the podium of the 1936 Olympics in Nazi Germany having won the 100 meters in 10.3 seconds, and three other gold medals. Later, Dean Cromwell, the assistant coach to Owen's team, would attribute this beautiful act of sporting defiance to a crude manifestation of biological destiny:

> The Negro excels in the events he does because he is closer to the primitive than the white man. It was not long ago that his ability to sprint and jump was a life-and-death matter to him in the jungle.

The power of that victory in front of a racist murderous regime is sadly undermined by the racism that belittles the achievement itself. Attitudes such as these are extremely common within sports and in the public. Matthew Huey and Devon Goss, two sociologists from the University of Connecticut, forensically took apart a century of attitudes to sporting success of black people, and found that a genetic advantage was a persistent theme.* Throughout the twentieth century, theories arose that attempted to explain the apparently disproportionately high presence of successful black sportsmen. The most persistent is the idea that black people have a higher proportion of "fast twitch" muscle fibers, a type of subcellular protein that is involved with explosive movement.

Of course *black* is virtually meaningless for the purposes of this argument. The genes that confer skin pigmentation are few, but mask a level of deeper genetic variation within Africa than without. That a Namibian and a Nigerian have more similar skin color than either do to a Swede masks the fact that the majority of their genes are more dissimilar to each other than they are to that same Swede. So if the main classifier is skin color, the differences that underlie dark skin are too great to support an argument of generic athletic superiority. We know, for example, that in parts of Africa, notably the highlands of Ethiopia, many long-standing populations have genetic adaptations to living at altitude. In that specific regard, these people are more genetically similar to Tibetans than to any other Africans, all of whom we would collectively typically describe as black. But this characteristic is unusual in sub-Saharan Africans

* Moreover, while success of black athletes was frequently put down to biology, success of white athletes was attributed to hard work and cognitive abilities. Interestingly, these views are not consistent over time though. In the first third of the twentieth century, the 5,000 meters was utterly dominated by Finnish athletes. A German writer named Jack Schumacher wrote that "Running is certainly in the blood of every Finn . . . [They] are like animals in the forest," as part of a diatribe of Aryan superiority.

in general. All other things being equal (which of course they never are), my genetic ability to process oxygen via a gene called ACE is no different from most Africans, though not many East Africans. Similarly, a particular version of the gene alpha-actinin-3, which is associated with the fast-twitch muscle fiber, is present in successful black sprinters, but is not exclusive to Africans, or indeed any particular regional or cultural group within Africa. A thorough 2014 review by the Brazilian sports scientist Rodrigo Vancini of the scientific literature on the genetics of African athletes concluded that the studies of the variation in these two genes, the ones most frequently associated with black sporting success, "do not fully explain the success of these athletes. It seems unlikely that Africa is producing unique genotypes that cannot be found in other parts of the world."

Part of the cultural argument is based around the possibility that slavery bred in these physical capabilities. The idea that underlies this assertion is that strength and physical prowess would be desirable in slaves, who would then be successful workers and thus procreate, and their genes passed on. This is a kind of "common sense" argument. But science is the opposite of common sense. It's a set of methodological tools that attempt to extract objective reality from how we perceive it. Science sets aside the bias that we lug around, and separates what feels right from what is.

There are several problems with the idea that slavery bred superhumans. The first is that 400 years is not enough time to establish particular alleles with that effect. Ten or twelve generations might provide the time for the spread (or eradication) of an allele of great biological significance. But as with so many human behaviors, we're not talking about a single gene of great impact. There are dozens of genes that are involved in the biology of sporting prowess, and these are not uniformly distributed across competitors of different sports: Sprinters do not make good long-distance runners. The second problem is that I am unaware of any data that has analyzed positive selection for these alleles in black people with slave ancestry. Without this, assertions of slavery being effectively a program of selective breeding are merely vaguely racist wish-fulfilment, confirmation bias, or yet another form of adaptationism.

Physical characteristics obviously do play essential roles in sporting success. The average height of an NBA basketball player is six feet seven inches, where being tall is obviously quite handy. Conversely, in horse racing, a sport dominated by white people, jockeys are typically small and light, broadly in accordance with Newtonian rules about speed and mass. Height is heavily influenced by genetics, but these numbers are nothing to do with race, as viewed through the lens of skin color or continental origin. The Dutch are the tallest people on average on Earth, and I have little doubt that if there were similar numbers of Dutch people as there are Americans, and basketball was as culturally important and ubiquitous, then they would produce teams as good as the LA Lakers.

Sport is sometimes cited as the great leveler, a forum in which only talent and sheer grit will win the day. The idea that black people are better at sport because of genetics, and possibly because of breeding during the wicked centuries of slavery, is built upon tissue foundations, and its cultural ubiquity yet another example of the chasm between what we think, and what science says is true.

Some look to the migration and history of our species as a means of suggesting and reinforcing a biological basis for the existence of race. There was, for a time, discussion about where modern humans originated. The question was whether we all spurted out of Africa as one species and grew into the current forms of *Homo sapiens* we see today, or whether earlier Homos had positioned themselves around the world in earlier migrations, and we have evolved into the current forms from those founding parents. This is known as the "multiregional hypothesis," but it has been almost universally rejected for many years. The bones don't say that; they show that the physical differences displayed in the living and the long dead are not significant enough to warrant classifying anyone living on Earth as having had a different route to the present. There are no physical or biological barriers to reproduction for anyone fertile on Earth either; an Aboriginal Australian could happily produce fertile children with an Aboriginal South American, or with anyone in Africa, despite the furthest genetic distance between them that we observe.

In the last couple of years, we've seen the addition of a nuance to the preeminently dominant Out of Africa hypothesis via the new ancient genetics discussed in Chapter 1. Modern humans clearly successfully interbred with both Neanderthals and Denisovans, and we carry their DNA to this day. These are not the emergence of other species of humans, but the incorporation of other species into our own. Their genetic contribution to us is not insignificant, and in some cases has furnished modern humans with some specific characteristics that otherwise we would not have. But these contributions are not enough, and not specific enough, to successfully feed into an argument that puts the broad colloquial definitions of race as separate evolutions. Mobility of human species and our excellence at sex have placed the common origin of all humans alive at only 3,400 years ago, or thereabouts. This means that the genetic bases of the characteristics that we class as racial are modern.

I believe that we don't have the language that allows us to align how we talk about race and what genetics and evolution has shown. Genetics has revealed that human variation and its distribution across the planet is more complex and demands more sophisticated squinting than any attempts to align it with crude and ill-defined terms like race, or even black, or white. It is for this reason that I am comfortable stating that from the point of view of a geneticist, race does not exist. It has no useful scientific value.

In science we crave precision, in measurement and in language. The urge to categorize is very human, and much sought after in science. We don't have a definition of life; we have inadequate definitions of species. Life does its best to undermine our noble attempts to categorize living and the living—this is what makes it exciting. Yes, variation that we see when measured by scrutiny of the genome broadly matches large landmasses, but even with oceans as barriers, these boundaries are not sharp, and still only account for a fraction of the differences between individuals. To establish these differences between people and peoples may yet have some value in terms of understanding diseases whose penetrance is unevenly distributed across all humans, and has focus on particular populations.

But again, this does not align with the popular concept of race. That, of course, does not mean the racism doesn't exist. It's a special thing to experience. Unlike millions, my life has not been blighted by persecution drawn from the tiny fraction of DNA that makes a person look different. But it is difficult to comprehend if you haven't experienced it. It was trivial in my case, but those experiences set a fire that burns and burns. Genetics has shown that the conflict is with people, and not embedded in biology.

Though Darwin did not discuss humans in *On the Origin of Species*, his second greatest work, *The Descent of Man*, was devoted to it. He had seen many of the indigenous populations of the world during his travels on the *Beagle*, and considered their physical traits carefully. He speaks of races and subspecies, using the language of the era, and embracing soft definitions that we no longer can support:

> But since he attained to the rank of manhood, he has diverged into distinct races, or as they may be more fitly called sub-species. Some of these, such as the Negro and European, are so distinct that, if specimens had been brought to a naturalist without any further information, they would undoubtedly have been considered by him as good and true species.

Darwin had used that emotive word *race* in *On the Origin of Species* to describe the varieties of types of living thing, "including the several races, for instance, of the cabbage." But central to his big idea was a recognition of the constant flow of creatures through time, and in 1871 in *The Descent of Man*, with great, possibly typical, prescience, he also acknowledged that those characters of race in manhood were neither permanent, nor quintessential:

> It may be doubted whether any character can be named which is distinctive of a race and is constant.

Not for the first time, genetics has confirmed what Darwin suspected. These were suspicions based on observations, with no concept of the mechanisms of inheritance that would follow him, many of which emerged from the work of his half cousin. Ironically, Galton enacted his own minor form of eugenics. He was married to Louisa Butler until the day he died, but they had no children, and his genetic makeup would not be passed on down the ages.

Galton's racial definitions are no longer defensible, and indeed none of the ways in which we talk about race today stands up to the scrutiny that genetics has enabled. Families are too untidy, human history is too convoluted, people too motile. The deck has been shuffled and reshuffled. Genetics has shown that people are different, and these differences cluster according to geography and culture, but never in a way that aligns with the traditional concepts of human races. Sometimes in this argument people might say "colors are just social constructs, but you can't deny that they exist." It is true that what we call "blue" is merely a convention for what we experience when visible light at a wavelength between 450 and 495 nanometers from the electromagnetic spectrum is processed first in our retinae and then brain. However, the electromagnetic spectrum is continuous, and these arbitrary markers are useful descriptions of what we experience. Human variation is continuous wherever we look too, but unlike light, not in a single line. People do lump together when we look at any single characteristic, and sometimes when we look at multiple traits. But we could equally look at other characteristics within the same groups and find different clustering patterns, and these are never fixed in a population, because populations are not fixed. The Jews once had high (but not exclusively high) rates of Tay-Sachs disease. Now they do not. Some Jews have ginger hair and pale skin, just like plenty of Scots. Others do not. The skin tone of the people of the Andaman Islands is very similar to that of the people of central Africa, but they acquired that hue via different historical and biological routes. Some black Africans are evolved to process oxygen at high altitudes, as are some Tibetans, but most are not. "Black" is no more a race than "long-distance runner" is.

We don't have to like people to accept that they were correct, or wrong, but Galton remains a tricky fish. Much of his science was utterly brilliant. Much of his insight was equally dazzling. Many of his opinions were horrid. It seems much of his motivation for doing science was born of these ugly views. Science is a process that strives to excise our limited view of the universe and our inbuilt prejudices from understanding an objective reality. Things are often not as they appear to us, but we invented and developed the scientific process to correct our subjective failings: Data is king. Francis Galton's inclination toward being a data junkie led him to instigate a science that he hoped would affirm his prejudices. The beautiful irony is that it did precisely the opposite.

6

The most wondrous map ever produced by humankind

"For every complicated problem there is a solution that is simple, direct, understandable, and wrong."

H. L. Mencken

May 2000, Cold Spring Harbor, NY

Every scientist knows that the most productive conversations happen in the bar. Science is a collaborative venture, and the idea of lone scientists beavering away on their life's work until that eureka moment is just a plain old myth. In research you hope that the data is true, but the interpretation of it requires discussion, debate, and argument, and so scientists go to conferences a lot, and present their work and argue about it and, when this game works, make it better. Yes, the presentations and formal lectures at scientific meetings can be dreadfully important, but—and this is a trade secret—they can be breathtakingly boring. Sometimes astonishing results are mired in slides so dense and incomprehensible that they make you want to poke the free conference pen in your eye, or strangle yourself with the name-tag lanyard. Sometimes slick, charismatic presentations mask thin, overstated, or dubious results. Sometimes, just staying awake is the biggest challenge.

But in the bar, the real science gets scrutinized and the best ideas assembled. Lifelong collaborations and friendships are made, bitter squabbles and permanent enmities are forged.* In May 2000, many of the best geneticists in the world had assembled at Cold Spring Harbor on the north shore of Long Island. They were there getting ready for the final furlong of the largest, grandest, most expensive project in the history of biology, and in the history of science, second only to the twenty-seven-mile-round particle accelerator at CERN that would later unveil the Higgs boson to the world. While the physicists were busying themselves trying to understand the fundamental structure of the universe, the geneticists were putting together the most difficult jigsaw puzzle ever attempted. The Human Genome Project had been running for eight years, and an end (though not "the" end) was in sight. Twenty-three chromosomes, made up of 3 billion letters of genetic code, the instructions to make a human being were the quarry.

It's always been described as a mapping project, and this is an excellent analogy, though the genome project was an inverted version of real cartography. Explorers have been charting rivers and coasts and mountains for millennia, and these maps got bigger and bigger over time, the scale increasing and the parts fitting together as we zoomed out from our immediate vicinity. Ultimately, we only saw the picture of how things really are in 1968 when *Apollo 8* astronaut Bill Anders took the first decent photo of the earth as it rose above the Moon. In that picture we saw the world for what it is. In the images of home that have been captured since, from the International Space Station, from satellites, and now so casually from software like Google Earth, we can see every river, every hillock and mountain, every forest, city, village, house, streetlight, and

* I once witnessed an actual fistfight between senior academics, though I will never say who, so don't ask. It didn't last long.

road. At night, the pictures taken from the Space Station show the glimmer of the city lights of every conurbation and illuminated trunk road, and the glows that flank the banks of the major rivers that like veins and arteries have fed civilization for all of human history. This is how the world fits together, and human civilization, trade, agriculture, and wars are carved out into the geology of our planet.

The genome is the other way round. Those classic images of chromosomes, the pinched paired socks all neatly lined up with Xs and Ys, those are the long-distance satellite images of the earth, taken from afar, with little resolution of the details, and almost nothing to say about the precision of a working cell, let alone a person. Down a microscope, they show the genetic continents and oceans, but nothing of the details. Genetics is all about the details. We knew some of the large-scale problems of DNA in the time before the genome project. A handful of cancers are caused by two chromosomes splitting asunder and rejoining in the wrong place. The breakage cuts a gene in half and renders it useless, and the protein that it normally encodes is one of many involved in controlling the rate of cell division. When it's broken, the cells divide uncontrollably, and that is a tumor. The particular genetic defect behind this type of cancer is visible from those low-resolution images, but all of the couple of hundred other cancers we work so hard to fix are coded into the fine details of the genome, still invisible under even the most powerful microscope.

Down syndrome is caused by having three copies of chromosome 21 instead of the normal pair. Turner's syndrome—a female disorder with a suite of physical and reproductive problems—is a result of a woman missing a second X. Klinefelter's syndrome is when a man has an extra X. The discoverers of XYY syndrome didn't get to attach their names to it, and that is what it is called. All of these are the equivalent of huge continental catastrophes that like an erupting volcano are visible from space. They're relatively, mercifully, rare. Almost every genetic disease (and normal trait) is hidden in the mountains, plains, houses, streets, and tributaries.

Mapping out every house, street, and tributary was the primary aim of the HGP. The genome is the total sum of all genetic material in an organism. Almost all of it is contained within the chromosomes, apart from the little rings of DNA that the cellular powerhouses mitochondria hang on to, this betraying their evolutionary origins as bacteria that were annexed into another cell some 2 billion years ago. In the human genome, in total, there are around 3 billion individual letters of DNA. Of the analogies of scale, the one that gets trotted out most frequently is that this is equivalent to some twenty standard-issue phone books, though when I use that in lectures these days most school kids have never seen a phone book.

The point is that it is a lot. By the time of that meeting at Cold Spring Harbor, we had identified thousands of genes, most through the slow, laborious processes pioneered by those who found the first disease genes in the late 1980s and early '90s—cystic fibrosis, Huntington's, Duchenne muscular dystrophy. But, even with the gold rush of gene identification that was continuing apace in the 1990s, there were still fundamental gaps in our knowledge about our own DNA. The biggest and most obvious of these was this: How many genes does a human have?

Back in the Cold Spring Harbor bar in 2000, a young British geneticist named Ewan Birney was fooling around, but inadvertently doing something quite profound at the same time. Nowadays it has become a tiresome cliché to say that a person's passion or quintessential characteristic is "in their DNA." The satirical magazine *Private Eye* has a whole column dedicated to this phrase flopping out of journalists' and celebrities' mouths. Well, Ewan

Birney is a man with DNA in his DNA. These days he heads the European Bioinformatics Institute in Hinxton, just outside Cambridge, one of the great global genome powerhouses. While our contemporaries went off to Koh Samui or Goa to find themselves on their year off before going up to university, Ewan had won a place in the lab of James Watson, at Cold Spring Harbor, just at the birth of genomics, the biological science that would come to dominate all others.

Maybe it was his familiarity with that bar—or maybe it was just the beer—that led him to do something quite silly, fairly trivial, but something in fact that is one of the great comments on the nature of science. Ewan opened a betting book one night in the bar. He pestered the leading geneticists in the world, all assembled at the meeting and, for a dollar a bet, asked them to predict how many genes a human has. The prize for the nearest guess was the pot and a bottle of Scotch.

Just like the genome itself, the page mutated over the course of the evening, and over the next couple of years. It's a scribbly mess, with crossings out, nine additional conditions, and five footnotes. They were added because, being scientists, these gamblers were all concerned with precise definitions, and there was disagreement and squabbling about what a gene actually is. So, along with some of the admin rules and housekeeping notes (contact details, when the decision would be taken, "no pencil bets," "one bet per person per year"), the footnotes specify what a gene is, as best they could agree at the time.

Everyone was in. Over the next three years, 460 geneticists parted with their cash (the stakes went up to $5 in 2001 and $20 in 2003). Reading it now is like the who's who of geneticists at the turn of the twentieth century. Names of scientific giants who we learned about as undergraduates, whose techniques were copied by

GENE SWEEPSTAKE 2000–2003

http://www.ensembl.org/genesweep.html

1. Costs $1 2000, $5 2001, $20 2002 to bet.
2. Bets are for one number, winner takes all
3. A gene is a set of connected exons by transcription/mRNA splicing + protein coding γ, β, α, *, †
4. Assessment of Gene number method will occur via agreement on 2002 CSHL meeting
5. Actual assessment of Gene number will occur on the 2003 CSHL Genome meeting (or equivalent)
6. Write your email, number, + payment in the book
7. One bet per person, per year
8. No pencil bets
9. Stay at CSHL. Contact David Stewart

stewart@cshl.org

α CDS in repetitive regions are not counted even if expressed.
γ Autosomal + X + Y chromosomes from the reference sequence. no mitochondrial genome.
* At least one transcript must encode a protein.
β Ig and T cell genes are only one gene per loci
† If transplicing exists, each trans splice is a separate gene

The gene sweepstake betting book.
Ewan Birney opened a betting book at a meeting of the Human Genome Project in 2000. For a dollar a bet, he asked the world's top geneticists to predict how many genes a human has. The prize for the nearest guess was the pot and a bottle of Scotch. The winning bet, by Lee Rowen, was 25,947. The real number is around 20,000.

all. Winners and some destined to win a Nobel Prize are in the book.* This was the single group of people who would be better qualified than anyone in history to answer this simple question, for science, and for a bottle of single malt.

All of them were wrong. And not just a little bit out. This was not a competition clinched by slight margins; most entries were thousands, some tens of thousands wrong. The highest prediction was from a British scientist named Paul Denny—a whopping 291,059; plenty were above 150,000 and many were around the 70,000 mark. Birney's entry is first: 48,251. As agreed in the book, the decision would be taken in 2003, using a method of counting genes as specified in the footnotes of the betting book. Back at Cold Spring Harbor, the jackpot eventually went to a Lee Rowen, then a forty-nine-year-old researcher who ran one of the first big genome centers under the leadership of Leroy Hood, then and now one of the leading geneticists in the world. She had been in the bar that night in 2000 when the definitions of a gene were being ironed out over beer. Her bet was 25,947.

The real number is around 20,000. This is the same number of pieces as the four largest Lego sets combined.† The best experts in the world were all mistaken.

Even the definition of a gene is not rock solid either. The betting book specified sections of DNA that coded a protein. Nowadays we know of many bits of DNA that only encode that supposed intermediary molecule RNA, and which never make it to the status of a

* Sir Richard Roberts earned his in 1993 for identifying the strange space fillers in genes called introns, which are discussed later in this chapter. John Sulston, one of the driving forces behind the British effort on the HGP got his call from Stockholm in 2003 for his research into cell death in the nematode worm.

† The Taj Mahal in Agra, the *Millennium Falcon* from *Star Wars*, London's Tower Bridge, and the Death Star, also from *Star Wars*.

protein, but nevertheless have essential biological functions. Are they genes? Well, probably, sort of. As ever, biology turned out to be more complex and interesting than we had imagined.

Lee Rowen got an envelope of cash, but never claimed the whisky. She told me that she didn't spend her winnings either: "How could I spend it when we know the number is wrong?" Being wrong is the backbone of science, and I am certainly not ridiculing these predictions. Ignorance is the position from which we work out what is correct. Ewan Birney's betting book is a great demonstration of how science works, and the virtues of ignorance. It showed that, at that time, we didn't really know how genetics works in humans. When talking about this era, Birney invokes a much-maligned sentiment expressed by an unlikely accidental philosopher: Donald Rumsfeld. In February 2002, the then US Secretary of Defense said this about the existence (or otherwise) of weapons of mass destruction in the second Iraq War:

> As we know, there are known knowns; there are things we know we know. We also know there are known unknowns; that is to say we know there are some things we do not know. But there are also unknown unknowns—the ones we don't know we don't know.

Nestled deep in all that wretched inelegance is great wisdom. The range and error of the betting book showed how clearly human genetics was set in the zone of unknown unknowns until the twenty-first century. We didn't know how many genes we had, and our model was roughly that there would be a gene that corresponded to every disease, to every trait. We are sophisticated creatures, but our biology is fundamentally no different from a chimpanzee or a cat. But we have enormous intellectual powers that dwarf any other creature. Dolphins, monkeys, crows, octopuses all display facets of intelligent life—problem solving, tool use, complex communication abilities. We may well laud those mad skills, but they are still light years away from us in every single one of those categories. And so it might not be unreasonable to assume that our own faculties would be encoded in a genome that reflected those powers, at least in terms of numbers.

But we don't have more protein-coding genes than a chimpanzee. In fact, we have fewer genes than a roundworm. Or a banana. Or *Daphnia*, a type of minuscule see-through water flea the size of a grain of rice. Or indeed a grain of rice.* We have roughly the same number as our most useful genetic test species, the mouse, and a few more than our second favorite lab rat, the fruit fly. Just consider what is happening right now: I'm typing these words with a manual dexterity unique to living organisms, and I'm conceiving a story with reference to memory, deep understanding, creativity, and an ability to imagine the future, in which you are using all the same faculties in reading them, you're imagining me typing right now. We estimate based on the number and density of connections between the neurons in our skulls that the brains you and I are using right now are the most complex objects in the known universe. Yet the code that underwrites that spectacular lump of gray meat is basically the same as animals that can do none of this.

The greatest achievement of the Human Genome Project was working out exactly how little we knew—known unknowns. Once you know what you need to know, the future is laid out in front of you. And so, the map was sketched, and the landscape was set out—where to explore, and what we might be hunting for.

* Plant genomics is even more weird and unpredictable than in animals. Many plants have enormous genomes, and we're not sure why, and many have multiple copies of chromosomes. They still run off the same basic and universal principles of biology that show an unequivocal shared ancestry, but when it comes to the mysteries of genomes, plants make ours look like Lego.

If the paucity of genes was the first great revelation of the Human Genome Project, the second was that almost all the genome is not genes at all. The exome—the DNA in a genome that encodes actual proteins that perform the jobs of living—constitutes less than 2 percent of the total amount of DNA that you carry. Imagine that in a novel: It's as if there were just 300 meaningful sentences in Dostoevsky's *Crime and Punishment*, and the rest of the 211,591 words were largely incomprehensible twaddle. Or in the book you are reading right now, just 150 interesting sentences. You be the judge of that.

We had known for a couple of decades that much of the genome did not specifically code for proteins. Much of it is dedicated to the architecture of the genome. DNA is a double helix, the iconic twisted ladder rising up and screwing to the right that Crick and Watson put together in 1953.

It is, however, almost never like that. DNA within a living cell is continuously busy and dynamic. It unwinds and rewinds, and gets edited and proofread all the time. During the cell cycle, when one cell fattens and divides into two, as every single cell in the last 4 billion years has, all of the DNA contained within it is duplicated, such that the daughter cell contains the same genome as the parent (except for sex cells, which end up in four daughter cells, each with half the genetic material). For that to happen, a carefully choreographed dance ensues, where loose strands untangle themselves and wind up into the chromosomes that we know so well. They're only in this form for a brief moment in a cell's normal life cycle, and to bundle up like that requires huge parts of the DNA to be concerned not with encoding proteins, but with scaffolding.

Then there's all the instructions on what genes do when and where. Every cell contains every gene, but will only require a handful of genes to be active at any one time. A gene that prompts cell division is no use in a tissue that has finished growing, otherwise it becomes a tumor. A gene that is active in sperm production is not required anywhere other than in a man's testes. There is a time and place for every gene, especially during the development of an embryo. I worked primarily on a gene called CHX10 (pronounced "chox ten"), which in mammals is involved in the process of growing an eye. It is active just after the basic shape of the eye is formed, in a mouse around ten days after conception, in a human around ten weeks. There, at that time, CHX10 sets up a program that tells the cells that will go on to form the retina to multiply. These are neurons, brain cells, and sit in between the outside of the eye and the lens. When CHX10 is broken, a child's eyes don't grow properly, and they are born blind, with shrunken deep-set eyes, in a condition called microphthalmia.

The protein that the gene CHX10 encodes doesn't have a role in the body other than to turn on other genes during those crucial stages of embryonic development. It isn't an enzyme that speeds up the digestion of food, or a protein that builds skin, bone, or hair, nor is it one that ferries oxygen around in the blood, or converts a photon into an electrical signal so that we can see. The CHX10 protein folds itself up into a ribbon with a groove on one side. The sole purpose of this opening (and by extension, the sole purpose of the CHX10 protein) is to clamp itself onto a stretch of DNA. It only will bind to a particular sequence of DNA just eight letters long—TAATTAGC. This is not a gene, but sits near one, normally before the beginning, and when CHX10 clamps on, it acts as an "on" switch. It belongs to a family of genes called "transcription factors," and everything they do is to regulate others by controlling these switches. They rule over cascades of genes, mostly during the development of the embryo, and issue generic instructions—"this area of cells is going to be an eye"; "this bit of the eye should be retina"; "this part of the retina should be rods and cones" or conversely "this part should not be photoreceptors." These cascades are enacted by other genes downstream of these master controllers, and at each stage the cells move away from their original state of having endless potential into a position of high specialization—"you're not merely a type of brain cell, you're a photoreceptor that only recognizes light in the wavelength that we call blue." Transcription factors are foremen of grand building projects, who look at the map, point to an area, and say "build something there." The transcription factors themselves are of course encoded in genes, but the switches they activate are not, and there are thousands of them scattered throughout the genome.

And the genes themselves are broken up by other bits of DNA, called introns, which don't encode proteins either. All human genes are punctuated with introns, and sometimes they are longer than the actual gene itself. It's a strange thing, to break up a working xxxxxxxxxx text with so many yyyyyy random bits of irrelevant zzzzz guff, and I continually find it impressive that a cell knows to edit it out when going from the basic code of DNA, via the temporary messenger version of the genetic code, RNA, to the fully functional protein.

And there are pseudogenes—they used to be active, but their function became unimportant in evolution, and they were at some point negatively selected. When they randomly mutated, as all DNA does, the outcome was negligible or nonexistent, and they are left to decompose in our genome. We know they once were important, because other animals still put them to good use. Whales, who can only smell when surfacing, have the remnants of hundreds of genes for smelling that dogs and mice still use. For us with our inurbane noses, plenty of olfactory receptor genes have nothing to add to our lives, but they are still there, slowly rusting in our genomes.

And then there are huge chunks of DNA that are just repeated sections. And then there are huge chunks of DNA that are just repeated sections. And then there are huge chunks of DNA that are just repeated sections. Many are repeated hundreds of times. Sometimes these repeats are of significance, as the number of repeats varies between people. These aren't the most glamorous regions of the genomes, but the unknowns within these repeats are awaiting discovery.

And there are huge chunks that don't do much at all, and we don't know why. These bits acquired the catchy name "junk DNA" in the 1960s, which has stuck like a curse ever since. It may just be filler, or bits that chromosomes have acquired and that have no effect. Not all non-coding DNA is junk, though, but all junk DNA is non-coding. Much is made of what junk DNA actually is, and whether it is really evolutionary junk shoved up in the attic that might one day be useful, or evolutionary rubbish that ought to go in the bin.

Part of the problem here is precisely this language. When we give technical bits of biology catchy names, they do have a habit of sticking, often beyond their usefulness within science. *Primeval soup* is one that stuck from the 1920s onward, to describe the hypothetical origin of life. The principle of just the right ingredients of chemicals coming together into a living broth is a fundamental mistake, and has hampered and steered that research area ever since and often in the wrong direction. Francis Crick named the core kernel of molecular biology the *central dogma* in 1956—the idea that DNA encodes RNA that translates into proteins. *Dogma* is a term that we science types have been trying to avoid since the seventeenth century, given that it means an incontrovertible belief laid down without evidence by an authority. It was puzzling that Crick should apply it here in the business of science, in an endeavor that relies exclusively on evidence and never on authority. The great historian of twentieth-century biology Horace Judson wrote that when quizzed on this oddness, Crick replied:

> I just didn't know what dogma meant. And I could just as well have called it the "Central Hypothesis" . . . Dogma was just a catch phrase.

Which just goes to show that the very best geniuses can also be idiots.* *Junk DNA* is one of those phrases too. It was coined by a Japanese American geneticist named Susumu Ohno,† and that simple analogy has meant that it draws much attention from the world outside gene jockeys. Years later, Ewan Birney, in his current role, led a gigantic international consortium to try to assess the role of the non-coding DNA of the human genome, including the stuff that might be called junk. It was a phenomenal project, and culminated in dozens of papers in 2012, and a huge database available to all as a resource for hunting for networks within our DNA. It was also the stuff of enthusiastic hype, too, including by the journal *Nature* where the main paper was published,‡ and by the press in general. The headline, nurtured by the papers themselves and the press releases that came alongside it, suggested that across the genome, some 80 percent of it had a biochemical function. They had been fishing for biological activity in the experiments, and those two things are not the same. How much of the non-coding genome is essential for human life is not accurately known.

---

* That's quite a good joke, and true, but in fact, Crick later clarified this apparent error, saying that however plausible the mechanism of the central dogma was, in 1957, there was little experimental evidence to support it. He said that he applied the word *dogma* with that in mind, as "all religious beliefs were without serious foundation."

† Ohno also tried to make music out of the sequences of real DNA encoding four notes of the normal octave scale. It's not very good.

‡ I was working at *Nature* at that time, and with Ewan, and as such I suppose was part of the hype engine. I wrote and produced an animation voiced by the actor/musician/comedian Tim Minchin, which was all part of the press surrounding the release of this dataset. It's a good cartoon, and I stand by it, while accepting that some of the coverage from the time was perhaps a bit breathless and overreaching. So it goes.

Reaction to the coverage was sometimes vitriolic, cruel, and quick. Some of it was directed at the science being faulty, in that DNA being chemistry in action was not the same as having a function in the body. Other scientists were outraged by the scale of the project—yet another example in the post-Human Genome Project era that only huge international collaborations could generate money and public interest, and little labs grinding away at unsexy projects were being ignored and underfunded. There may be some truth in that. The politics of science are often as complex and confusing as the data itself.

Part of the problem here is language. The coding proteins account for less than 2 percent of the total amount of the human genome. Lots of the rest of it does something, including the switches—the ons and offs for genes to dance their choreography as we develop in the womb, enact our lives, and interact with the rest of the universe. Some of it does stuff that we haven't discovered yet. Is it junk? No. Is it useful? We don't know. Most of the genome—upward of 85 percent—does not appear to be under any selective pressure at all. Many writers have described the non-coding realm of our DNA as the "dark matter of the genome," alluding to the stuff we know exists in space, that makes up the majority of mass in the universe, but that we cannot yet account for. We don't know what it is, but we infer that it is there because of our model of how the universe is built. I intensely dislike this phrase. Metaphors in science should clarify or enlighten, not obfuscate because they sound profound. To me, it is using one thing we don't understand to explain another, and thus has no explicatory power itself. Instead it merely reinforces the mystery, as if it were not simply a scientific problem of known unknowns, but something mystical. We have no room for the mystical in science.

❧

All these bits and bobs of genetic switches, scaffold, detritus, and mystery make up almost all of the genome. The Human Genome Project wasn't easy, and took the best part of a decade, the invention of brand new technologies, unprecedented computing power, and $3 billion. Here is why.

Imagine, if you will, that this very sentence is a gene.

It has a structure to it, and each word is important, though some (for example, the sub-clause *if you will*) are less important than others, and not entirely necessary to the overall meaning. The imperative verb is essential to the intended meaning, and so are the nouns—without *gene*, *sentence*, or *imagine*, the meaning is either changed or lost completely. Language is often used as an analogy for DNA and genes, obviously because both are composed of an alphabet, in which the order of the letters is essential to the intended meaning. DNA is a coded alphabet to be translated by the mechanics of a living cell into a protein; all life is made of, or by, proteins. So the analogy of language and genes works neatly, but only up to a certain extent, and not least because the meaning of these sentences is (with luck) right here on the page. With DNA, the meaning is coded, but much worse, it's not written out into our genomes in a very straightforward way. Evolution, blind and slow, has not inched along over billions of years with any intention that it should be decipherable to one or any of its billions of children.

In English, we put spaces between the words so we can read them easily, but in DNA punctuation is not visible. So it becomes:

Imagineifyouwillthatthisverysentenceisagene

In the genome, it doesn't sit on its own in a discrete sentence. Genes reside on chromosomes, punctuated by the apparently random introns mentioned earlier, and the points of insertions bear no relation to the sentence structure or meaning:

Imag ineify ouwillthat thisverysentenceisag ene

These bits that convey the meaning of the sentence are the exons—in DNA the code that will translate into a meaningful protein. Introns and exons are made up of the same letters in DNA, or in my example twenty-six letters of the English alphabet. Introns can be any length, typically a thousand letters. Here I'll keep it simple and just make them thirty letters long. They're mostly random, but also contain the annotation that specifies where the breaks are. I'm adopting STOP and START so we can see where the coding DNA ends and the intron begins and ends. It now becomes

Imag*STOP*ANSJTUWIRNASHTPQLESNI*START***ineifyou willthat***STOP*NJGUTHRBERTGOPLAMNSD*START***thisvery sentenceisag***STOP*RITUEYRHTFPLMNASCHJWS*START***ene**

There's also nonsense padding at the beginning and end. In the stuff in front of the beginning of the gene, there's often an instruction that it's coming up, such as the binding site that CHX10 will clamp onto in order to switch it on. Again reduced before we lose our collective minds, I've included just thirty, and my instruction I'm writing as SENTENCE COMING, followed by GO to indicate where the gene actually begins:

JVNFKJVFJVNLKN*SENTENCECOMING*laksmingshqwuing *GO***Imag***STOP*ANSJTUWIRNASHTPQLESNI*START***ineifyou willthat***STOP*NJGUTHRBERTGOPLAMNSD*START***thisvery sentenceisag***STOP*RITUEYRHTFPLMNASCHJWS*START* **ene**OSHFNDBUBVLSJFBJNBFKLSBKKFJBKJBNV

I've kept the original sentence in bold and in lower case, so we can still see it, and the specific instructions in italic upper case. But genes are not annotated like that. In the genome, every letter is weighted exactly the same as every other one. So it becomes:

JVNFKJVFJVNLKNSENTENCECOMINGLAKSMINGSH
QWUINGGOIMAGSTOPANSJTUWIRNASHTPQLESNIS
TARTINEIFYOUWILLTHATSTOPNJGUTHRBERTGO
PLAMNSDSTARTTHISVERYSENTENCEISAGSTOP
RITUEYRHTFPLMNASCHJWSSTARTENEOSHFNDBUB
VLSJFBJNBFKLSBKKFJBKJBNV

. . . which is pretty murky. And gives us an indication of why reading genomes is such a chore. That is 215 characters, vastly simplified and shortened to fit on a page, and encompassing the 26 letters of our alphabet. Here's a real bit of a gene:

ATGACGGGGAAAGCAGGGGAAGCGCTGAGCAAGC
CCAAAT CCGAGACAGTGGCCAAGAGTACCTCGGGGG
GCGCCCCGGCCAGGTGCACTGGGTTCGGCATCCAG
GAGATCCTGGGCTTGAACAAGGAGCCCCCGAGCTC
CCACCCGCGGGCAGCGCTCGACGGCCTGGCCCCCG
GGCACTTGCTGGCGGCGCGCTCAGTGCTCAGC
CCCGCGGGGGTGGGCGGCATGGGGCTTCTGGGGC
CCGGGGGGCTCCCTGGCTTCTACACGCAGCCCACC
TTCCTGGAAGTGCTGTCCGACCCGCAGAGCGTC
CACTTGCAGCCATTGGGCAGAGCATCGGGGCCGCTG
GACACCAGCCAGACGGCCAGCTCGGATTCTGAAGAT
GTTTCCTCCAGCGATCGAAAAATGTCCAAATCT
GCTTTAAACCAGACCAAGAAACGGAAGAAGCGGCGA
CACAGGACAATCTTTACCTCCTACCAGCTAGAG
GAGCTGGAGAAGGCATTCAACGAAGCCCACTACCCAG
ACGTCTATGCCCGGGAGATGCTGGCCATGAAAACG
GAGCTGCCGGAAGACAGGATACAGGTCTGGTTC
CAGAACCGTCGAGCCAAGTGGAGGAAGCGGGAGAAGT
GCTGGGGCCGGAGCAGTGTCATGGCGGAGTATGG
GCTCTACGGGGCCATGGTGCGGCACTCCATCCCCCT

GCCCGAGTCCATCCTCAAGTCAGCCAAGGATGGCAT
CATGGACTCCTGTGCCCCGTGGCTACTGGGATGCA
CAAAAAGTCGCTGGAGGCAGCAGCCGAGTCGGGGAG
GAAGCCCGAGGGGGAACGCCAGGCCCTGC
CCAAGCTCGACAAGATGGAGCAGGACGAGCGGGGC
CCCGACGCTCAGGCGGCCATCTCCCAGGAGGAACT
GAGGGAGAACAGCATTGCGGTGCTCCGGGC
CAAAGCTCAGGAGCACAGCACCAAAGTGCTGGGACT
GTGTCTGGGCCGGACAGCCTGGCCCGGAGTAC
CGAGAAGCCAGAGGAGGAGGAGGCCATGGATGAAGA
CAGGCCGGCGGAGAGGCTCAGTCCACCGCAGCTG
GAGGCATGGCTTAG

That's 1,086 letters long. Multiply that by 3 million and you have our genome. That particular sequence is a tiny bit of CHX10 from chromosome 14, in fact, the bit that encodes the DNA clamp mentioned earlier. There's only the four letters of DNA, and this is only the coding sequence, without the random introns, which do nothing other than break up the gene. You will note that even like that, it's incredibly impenetrable in that form, and desperately uninteresting. And though it most certainly does contain a code, it is impossible to work out what the code is unless you already know it; there are no inherent patterns that are decipherable. Fortunately, the genetic code had been cracked earlier in the twentieth century experimentally, not by code crunching (though the structure of the code had been speculated about as soon as Crick and Watson had unraveled the double helix structure; the Russian nuclear physicist George Gamow* had proposed a code in a typo-ridden and slightly bonkers letter to Crick in 1953. The code was wrong, but it

* Gamow was a brilliant and influential physicist, whose work laid significant foundations for Big Bang cosmology. He also wrote a paper with his student Ralph Alpher, and added his friend Hans Bethe to the author list for the sole

triggered something in Crick that ultimately led him to think correctly about how the code really worked).

There are around 20,000 genes like that sequence above, in all sorts of sizes, with all sorts of different-sized introns, hidden in 3 billion letters of human DNA. The gene sits within a chromosome somewhere in the genome. We have twenty-three pairs of chromosomes, one of each pair inherited from each parent. Each chromosome contains thousands of genes, but much more filler. In 1994, when Lee Rowen deposited the first sequenced section of human DNA bigger than 500,000 letters into public databases, it was such a hefty piece of data that it crashed users' computers.

The full scale of the task ahead was colossal. Until computing power became sufficient to seek meaningful bits of DNA in the genome, this was a job that made hunting in haystacks for needles look like fun. Imagine a field with fifty rows of hay, each row a hundred meters long. Now imagine 600,000 of those fields. Now look for 20,000 needles in those fields. Except the needles are not metal, they're made of hay as well. And each needle is broken up into pieces.

But that is what they did. In the mid-1990s the technology was being developed at a feverish pace, as it still is today, and new techniques to improve and speed up genome sequencing were introduced. The one that came to dominate is called shotgun sequencing, and involves blasting multiple copies of long bits of DNA into thousands of smaller bits and sequencing them all. Reading shorter sequences is easier, so by generating copies of random fragments, you generate enough overlap between them all to get a full read of the original stretch of DNA.

---

reason that it would read "Alpher Bethe Gamow," and is known as the "aßy paper." He tried unsuccessfully to escape Soviet Russia in 1932 in a canoe. Twice. I thoroughly recommend *Life's Greatest Secret* by Matthew Cobb for the definitive version of the story of how we understand the genetic code.

❧

On June 26, 2000, the political machinations of big science ground into gear to plant their flag in the ground. The first map of the human genome was complete. President Bill Clinton stood on a stage in the White House East Room in front of the world's press, flanked by the leaders of the publicly funded Human Genome Project, Francis Collins, and of the private version of the same endeavor, Craig Venter. Behind them, on a live video screen, was the bobbing head of Prime Minister Tony Blair, representing the significant British involvement in the project. Clinton described the first human genome as

> the most wondrous map ever produced by humankind . . . today we are learning the language in which God created life.

Whatever your religious sensibilities, that was a bold statement. Of course, it's not quite right either, regardless of your religious sensibilities. As they had indeed discovered, the vast majority of the genome is not a language that can be translated at all. Only the genes carry any sort of coded meaning, and the rest is of some use, or not at all. We had already cracked the language of genes decades earlier. This was more akin to sorting out God's massive and extremely chaotic filing cabinet. But of course that doesn't quite have the same ring to it.

I remember that day very clearly. I wasn't trying to learn a divine language. It coincided with a moment during my PhD when I was trying to find the gene CHX10 in the databases of the so-called complete human genome. I could see where most of it was, in a section on chromosome 14, but it was far from complete, and it was unknown which direction it was pointing in. Unlike sentences in books, which run from left to right in English from the first word to the end, genes can point in either direction. .noitcerid rehtie ni tniop nac seneg ,dne eht ot drow tsrif eht morf hsilgnE ni thgir ot tfel morf nur hcihw ,skoob ni secnetnes ekilnU Thousands of researchers around the world were in the same position, stalking their gene, or looking for bits of DNA that their gene interacts with. The databases provided clues to the presence of genes yet to be discovered, but back in 2000 they were incomplete and muddled. I found one that was very similar to CHX10 on a different chromosome, and spent some serious time trying to find its beginning and end, and trying to work out what its purpose might be. Then one day, a research paper from another lab landed on my desk that described it in full detail. So it goes.

If this was the language of God, then a holy editor would've been a great asset. It was a huge grandstanding event, and I suppose some presidential showboating was appropriate; maybe science needs it every so often to get the public (who fund this work) excited and on board. But the genome wasn't complete. In fact, it was almost comically incomplete. This version was a rough draft at best. It wasn't complete in February 2001, when *Nature* published the formal paper of the Human Genome Project. It wasn't complete in April 2003, when it was officially declared finished with a database of 99 percent of the gene-rich DNA issued. Nor in May the next year when a quality control study confirmed this accuracy, though only just over 92 percent of the coverage. Today, the Human Genome Project lives on as a database, a reference sequence that is updated once in a while.

The tools and techniques that fell out of the HGP have built on those foundations to provide the subsequent stages in the exploration of our DNA. HapMap was the next big genome project, which set out to scan the DNA of people all over the world to identify the individual differences between large groups of people, and the genetic variation among them that might increase an individual's risk of a specific disease or characteristic. Language, books, and words are often used as analogies for DNA, and with good reason.

Letters make words, words make sentences, sentences make paragraphs, and so on. Typos and edits can change meanings subtly (inequity and iniquity), or profoundly (weak becomes week, and all sense is lost). Or meaning can be reversed: "I'm not ready" shifts to an opposite state of preparedness if the *t* in *not* becomes a *w*. Or single changes can have no effect at all: *cozy* and *cosy* are equally snug. Most books don't change that much between printings, and most have short lives, destined to be shelved in libraries and thus preserved unchanging. That earlier version of the language of God, the Bible, is a useful counterexample because it has been translated and changed and retranslated and reinterpreted endless times over thousands of years. Because it carries important cultural messages and information, sometimes beautifully written, it is revered and studied. The precision in those translations, like in genetics, is paramount.

For example: In early versions of the Book of Isaiah, written in Hebrew, there is a prophecy that uses the word *almah* to describe the mother of a boy named Immanuel, meaning "God is with us." Almah has no direct translation in English, nor in ancient Greek, but broadly means "young woman," or "woman who has not yet borne a child." By the time of Jesus, the Jews had adopted Greek and Aramaic, and no longer spoke Hebrew. Almah became the Greek *Parthenos*, which has a more specific meaning as "virgin." It's the root of a good biological term, parthenogenesis, used to describe the generation of young in some insects and reptiles in the absence of a male: a virgin birth. But in a mutated translation of a single word, the woman becomes a virgin, and the child becomes the Messiah, and the story of Jesus has instantaneously been transformed. Matthew and Luke render it true in the New Testament, a billion Catholics hold this as gospel, and that's what we all sing in Christmas carols.

Edits can be less subtle, and more like biology. Endless imperfect copying is essential in DNA, otherwise generations of an organism are not fit to adapt to the changing environment. In the written word, mutation might not be that desirable, especially if those words are Commandments. In 1631, the royal printers in London published a revised version of the King James Bible. In Exodus 20:14, a single word was omitted, whether by accident or satanic diktat is unknown, and that word was *not*. The book became known as the "Wicked Bible," as Commandment number seven in this verse decrees that "Thou shalt commit adultery." That mutation didn't survive very long, after some hasty embarrassed book burning.

These mutations in DNA are called polymorphisms, and over the years, the HGP and HapMap were designed to find as many of the typos and spelling variations in the human genome as possible. All of the dozens of versions of the Bible currently available are still Bibles, and they all remain the tales and lessons of God and Jesus and that sort of thing. And so it is with genetics. There is no suggestion that any person on Earth is not a *Homo sapiens*. It is sometimes said that humans are all 99.9 percent the same at the level of DNA. It's a slightly meaningless, glib factoid, as it doesn't make any comment on the nature of those differences, but broadly it means that our DNA varies on average, by one part in a thousand. From a sheer numbers point of view this is still a lot. We have 3 billion letters of genetic code. One thousandth of that is still 3 million individual points of variation. And that's just the SNPs.

Many differences between people are seen in chunks of DNA, which vary in the size of that chunk, and the number of times it is repeated. The number of copies of a section of DNA varies enormously from person to person. Very small repeating units are called microsatellites—they are normally between five and ten bases long, and can be repeated anywhere between five and fifty times. Minisatellites are typically ten to sixty bases long, and can be involved in how genes are regulated and how strongly they are turned on (see Chapter 7, on the horribly misnamed "warrior gene" MAOA), and have been useful in the type of DNA fingerprinting that has proved essential in the forensics of crime scenes. The studies of what these weird repeating units are for and why they evolved is ongoing. As far as a so-called complete draft of the human genome back in 2000, effectively none of this natural and important variation had been accounted for.

If this all sounds critical, in my opinion the HGP was one of the greatest scientific endeavors ever undertaken. It worked, and came in under budget, and on time. It was a visionary idea, a public project paid for by you, for all humankind. The sequences are freely available to everyone (with an Internet connection) in perpetuity. James Watson, John Sulston, and Francis Collins and a host of other top geneticists had seen the vision ahead as the only sensible way to proceed. The biomedical charity the Wellcome Trust knew it too and fought hard for funding a cohesive plan that would ultimately deliver the foundations on which biology would be built in the twenty-first century. The HGP fundamentally changed the way biology is done: Huge international collaborations are now normal and expected in science. Projects that include many disciplines, and many organisms, are the bases of research that delves into diseases, basic biology, evolution, and therapies, and all are underwritten by this colossal DNA database that was created. With hindsight, it is difficult to conceive of a way that biological science could have continued successfully without the Human Genome Project.*

* Craig Venter's privately funded project ran in parallel to the public consortium of the HGP, under the guise of a company called Celera. They both developed technologies that drove each other forward and spurred each other

That was fifteen years ago. Since then the amount of gene sequencing has increased exponentially as the cost of doing it has dramatically fallen. We know almost all of the genes now, and where they are. We know literally millions of subtle differences between genes in different people. We know how many single genes have multiple outputs via alternative splicing: A long transcript in RNA of the genomic DNA is nipped and tucked to produce proteins with very different functions. Yet we still don't know what much of the genome is doing. On that June day in 2000, President Clinton also pronounced that the end of disease was just around the corner.

> We are gaining ever more awe for the complexity, the beauty, the wonder of God's most divine and sacred gift. With this profound new knowledge, humankind is on the verge of gaining immense, new power to heal. Genome science will have a real impact on all our lives—and even more, on the lives of our children. It will revolutionize the diagnosis, prevention, and treatment of most, if not all, human diseases.

It is true that medicine has been revolutionized in the years following the HGP. We understand more about the causes of diseases than at any time in history. We know how diseases vary within and between populations. Genetic diagnoses of cancers can show precisely what type of cancer it is, and therefore how likely it is to react to different treatment regimes. We know that the genomes of

---

on in what was effectively a race. It was technically a draw. The journal *Science* published Venter's completed work, which included, it turned out, his own genome sequence, and *Nature* published the HGP's results. They colluded to publish on the same day and announce together. The HGP has effectively become the benchmark for genomic data, and is by design freely available to all, in perpetuity. It is one of the most highly cited academic papers in the history of academia. I have not met anyone who has accessed the Celera genome for reasons other than curiosity. It is literally a footnote here and, in my opinion, a footnote in the history of science.

cancers change and evolve as the tumors grow, making medicine's quarry even harder to tackle, but potentially offering up new and highly personalized treatments. However, the number of diseases that have been eradicated as a result of our knowing the genome? Zero. The number of diseases that have been cured as a result of gene therapy? Zero.

That wasn't its point. The genome is data, and science is built on data. Nowadays, DNA is used routinely in the diagnosis of dozens of cancers, of heart arrhythmias, in identifying the causes of thousands of diseases too rare to have historically warranted major research projects. The expectations of such a grand endeavor were huge, and perhaps misplaced. What we started uncovering in the wake of the HGP was that complex characteristics and complex diseases are affected by many genes, and, within those genes, variations that look innocuous enough, but cumulatively might correspond with a syndrome or behavior.

Even those first, almost obvious genetic disorders such as cystic fibrosis turned out to be not as straightforward as imagined in the 1980s. It's a terrible disease that affects around 70,000 people worldwide, causing lungs and airways to be blocked with mucus, and great difficulty in breathing as a result. Although treatments have profoundly improved the quality and survival of CF patients, life span remains severely truncated, and survival into one's fifties is a rarity. We'd known for decades that CF is a genetic disease, passed down through families in a very predictable way, and since the 1980s the gene responsible was identified as CFTR: When it's short, as in most CF patients, or otherwise broken, lung function is severely curtailed. It is an autosomal recessive disease, meaning that the faulty gene is on one of the chromosomes that doesn't relate to sex, and to get the disease you have to have two copies of that disrupted gene. We knew all that in the age well before gene sequencing and the Human Genome Project, mostly by watching the patterns in family trees. Carriers of a single defective CFTR gene have no symptoms, but if two carriers mate, then each child conceived has a one in four chance of inheriting both disease genes, and will be born with cystic fibrosis.* The pattern is perfectly Mendelian, and in the 1980s, that meant it was a good candidate for geneticists to focus on: a disease with a clear inheritance pattern.

Around three quarters of patients in the West have the same mutation, but there is a range of how badly the disease manifests itself. Not all patients with the same mutation have the same symptoms, which we might have once put down to random variation, or just unaccountable differences between people. In autumn 2015, with research still trundling along trying to unlock the mysteries of the longest-running attempts to understand a genetic disease, a set of five genetic modifiers were identified that were not part of the CFTR gene but changed the severity of symptoms. How these genetic variations affect the patients isn't known; they're not on the same chromosome as CFTR, which is also not unusual in our genomes. Chromosomes are not randomly assigned genetic storage units. The number of chromosomes a species has is fixed, or at least in humans has been fixed for millions of years, and we inherit one from each parent, each containing matching genes. So we inherit big chunks of DNA together, rather like clumsy shuffling in a deck of cards. But there's certainly no logical design behind the structure of chromosomes, and the utility of genes across a whole genome is in no way reflected by their arrangement on the twenty-three pairs of chromosomes. That we now know many of these genetic pressure groups (and there will be more) that exert power over this particular gene is both surprising and not, at the same time.

---

* That doesn't mean that if you have four children, the odds are that one of them will be affected. Each child has the same probability, as each conception is an independent event. In exactly the same way, the probability that you will win any prize in the lottery remains the same if you enter each week.

It's surprising as CF is a disease we know a lot about, having studied it since the beginning of the modern era of genetics, years before the HGP was even conceived. It's equally unsurprising, as this is what all genetics is beginning to look like in the era of the genome. Even the few genes that do seem to have a single clear function, the ones that can be tracked though families like peas in a Mendelian pod, don't operate alone, and aren't isolated from the milieu of human biology. Instead the influences of other genes can be felt, sometimes profoundly and from afar, at least from anywhere within our own genome.

This is one minor study among thousands. How we know this about cystic fibrosis comes from a technique invented in those first few years of the post-genome age. The standard techniques for identifying genes and their effects had been to look in people who had a disease and try to isolate the root cause. Those first disease genes were found by taking families with the disease and meandering through their genomes to find the bits that were broadly the same, and like a military pincer formation marching toward the target until it was met, and then sequenced. In the 1980s and '90s this was a slow process, though the HGP gold rush on genes made it much easier.

After sequencing DNA became cheap and easy, we started to do something different: Take as many people as you can find who have the same condition, and then look through all of their DNA to see if anything leaps out as being the same in them, and not in people without that disease. So in the case of the cystic fibrosis modifiers, the researchers took the genomes of 6,000 patients and scanned them at thousands of individual sites of variation to see if any stood out. The answers, after some juicy statistics, were five significant variants. These are simply bits of DNA that are different in people with the disease.

The technique is called a "genome-wide association study," universally referred to as a GWAS (and pronounced gee-woz). The points of variation don't necessarily tell you anything about the cause, only that there is something unusual going on at that spot in the genome. Thousands of people, thousands of spots of natural genetic variation, spread over twenty-three chromosomes. If you took two people and compared their genomes at these points you'd see lots of differences. If you take a thousand, you see more differences, but they average out into the overall morass of genetic variation in a population. But if those thousand people all have a shared characteristic, then some of the differences leap (or more often inch) out of the sludge.*

GWAS studies are represented graphically on what are known as "Manhattan plots," because they resemble a city skyline. If your sample is people without any specific shared trait, then you'll get twenty-three city blocks, each one a chromosome, comprising thousands of individual points of variation forming buildings all roughly the same height. But if they all have something in common, such as cystic fibrosis, then all of sudden a few skyscrapers start poking into the skyline. Where the base of a peak lies in your DNA might be part of the underlying problem. These are "association" studies, so don't identify causes per se. If you get lucky, you might do a GWAS and the big peak will reside in a part of the genome that is known, and it sits within a known gene, and the peak represents a variation that we can show triggers that particular defect. Returning to the concept of the genome as a map of the world, this would be like sticking a flag in the ground and saying not only that X

---

* They only do so after testing the probability that the variation is likely to be associated with the disease and not some random variation; this is done by comparing the "odds ratios," which is a statistical tool for establishing whether and how strong the occurrence of one thing is connected with the occurrence of another. In the case of GWAS, it's used to measure if the presence of a particular genotype is genuinely linked to the disease in question.

marks the spot of a problem, but we've sent in some investigators to that landmark and they have worked out what the cause is, say, a tectonic fault line, or a poisoned stream. But more often than not, GWAS studies return a map with a few, dozens, or even hundreds of flags planted in the ground, and when we send in the investigators, what they find is either nothing of particular note, or lots of small issues that apparently have a cumulative effect.

The first GWAS was published in 2005, when the genomes of just ninety-three people, all with a disease of the retina, were scanned and a section on chromosome 1 shone out. They found the gene that harbored this variant, and upon careful statistical analysis, showed that around 43 percent of people with age-related macular degeneration had this one particular natural genetic typo. Since that first study, the numbers and the power of this technique have skyrocketed. The real breakthrough came in 2007, with a gargantuan study called the Wellcome Trust Case Control Consortium: 50 labs; 17,000 patients with seven key disorders (coronary heart disease, diabetes types 1 and 2, hypertension, bipolar, Crohn's, and rheumatoid arthritis); 500,000 SNPs tested; and hundreds of new genes were found to have small but significant effects in these common diseases.

More than anything, this study set the bar for the future. Infinite human variation will be revealed with ever bigger samples sizes, huge numbers of people and patients, deep data. With greater data, we would be able to tease out the fine, finer, finest details of our genetic makeup. In May 2016, one study sampled over 300,000 people looking for genetic markers that associated with how long people stay in formal education. They found 74 genetic locations of statistical significance. They only account for a very small proportion of the reasons why people stay in school, but alongside motivation, parenting, intelligence, and a host of other factors, DNA is one.

Genomes became digitized on chips, where thousands of naturally occurring variations were logged onto small glass plates on which the DNA of your test subjects could be washed over. Where the DNA was the same, they would stick together, and via some simple chemical flags, would glow. That way, we could easily pick out the natural differences in our genomes that associated with any particular condition we wished to test. Hundreds of GWAS papers are published every year nowadays. Eleven years after ninety-three patients were the subject of the very first, the latest GWAS on age-related macular degeneration was published in January 2016, by a huge multinational syndicate: They analyzed the genomes of more than 16,000 patients and found 52 variants at 34 different positions in the genome.

There's been plenty of criticism of these studies over the years, some technical, some of poor experimental design facilitated by the ease with which they can be deployed, with gene chips a standard part of most human genetics labs these days. But probably the greatest intellectual contribution that GWAS has made over the years is that our assumptions, yet again, were proved wrong. They uncovered the biggest current mystery in genetics, one that we didn't know existed.

The mystery of the missing heritability

Many had thought that, broadly, common diseases would have common genetic causes. That assumption was born out of a misunderstanding similar to the one that made us wager on a vastly inflated number of genes at the inception of the genomic age.

It didn't turn out that way. So many of the genetic skylines that emerged out of the ongoing era of the GWAS did not come out as the high clear peaks of Manhattan, or even London, but the low skyline of Oxford or Cambridge.

Diseases that we know from family histories and twin studies that have a strong measurable heritable component gave up few major skyscrapers that revealed a key faulty gene. Instead, we got dozens or even hundreds of small peaks, many not qualifying as being statistically robust alone, but frequent enough to stand out a little. Many of the variations seen in people with a disease under investigation were known to us, but appeared to play no biologically relevant role in the disease etiology. It was as if you were reading an article, and you disagree with the general premise, but can't quite identify one particular error. The overall sense is wrong, but the precise mistakes are elusive and well disguised.

GWAS studies caused us all to scratch our heads. We weren't seeing the genetic roots of diseases that we were expecting from a century of work that started, as so many of the tales of human genetics do, with Francis Galton. He had the idea to use nature's freakish clones to study inheritance—identical twins. Twins come in two types: monozygotic and dizygotic. Monozygotic twins are identical, as very early in development following fertilization, the embryo divides but separates, leaving two genetically identical bundles of cells, each destined to grow into a child. Dizygotic twins are the result of two eggs being fertilized simultaneously, and so are no more genetically similar than siblings; they just happen to share a womb.

In 1874, Francis Galton sent out questionnaires to families and superintendents of various hospitals in order to

> collect data for estimating the respective shares that "Nature" and "Nurture" ordinarily contribute to the body and mind of adults, meaning by "nature" everything that is inborn, and by "nurture," every influence subsequent to birth.

. . . and got back ninety-four responses concerning twins. He didn't quite get the mechanism right (these were the days before molecular genetics) but he had spotted that, in principle, identical twins have the same inborn (that is, genetic) material as each other, and so any measurable differences between them could be accounted for as environmental rather than intrinsic. This simple premise is the foundation of much contemporary genetics.

With these comparisons, Galton was establishing a pattern of science designed to partition what was genetic and what was environmental. Genes are determined at conception, and are in principle fixed from that point on. He was building toward a concept that we have argued over ever since, called heritability. It's a tricky idea, and also a hideously named thing, because it sounds a bit like "inherited," which has a normal everyday meaning, and much of the confusion about what heritability means scientifically comes from this annoying scientific confusion.* Heritability is a measure of how much of the differences we see in a population can be accounted for by genetics, and how much is determined by the environment. Simple enough to say, but baffling once we look under the hood.

Let's assume that all Japanese people are born with black hair. That means that there is no variation in Japan for hair color, which means that the heritability of hair color in Japan is zero. It is an entirely genetically determined trait, but because all people have the same genetic makeup for hair color, there is no natural variation (at least until it goes gray, but let's ignore that for the sake of simplicity). Any hair color variation in Japan is a result of dyeing it, which is obviously not genetic.

---

* One researcher once told me that she wished that "heritability" was called something entirely made up—she suggested Tralfamadorian Score after the alien species in Kurt Vonnegut's *Slaughterhouse-Five*—and thus we wouldn't endlessly get tied into debates about what heritability actually means. The same might be said for the following genetic terminology: allele, microsatellite, linkage disequilibrium, epigenetics.

Sex is genetically determined at conception by the type of sperm that penetrates the egg. As there are two sexes in humans,* the heritability of sex is 100 percent—all variation we see in a population is determined by genetics. Or try this example: All humans speak a language (for simplicity's sake), which indicates that the capacity for language is genetically determined; *which* language you speak is entirely determined by where you are born, so it is environmental. So, there is no genetic influence on *what* language you speak, which means that all the diversity of languages spoken is environmental, and so its heritability is zero. To further confuse matters, heritability only refers to populations. It measures the differences seen across groups of people, and does not account for the amount of nature and nurture seen in an individual. There is no known measure for that ratio.

The examples above are absurdly simplified behaviors to attempt to clarify this hugely difficult metric. Most characteristics are a bit of both. Galton gave us the phrase *nature versus nurture*, but we know now that these are not in conflict. The complex interplay between nature—that is, genes—and nurture—that is, everything apart from genes—is how a person is built. *Nature via nurture* is a much better way of phrasing it.† Twin studies, over the years, have given us reams of data on heritability of complex human traits. There are none more complex than intelligence, and twins have shown us, fairly consistently over a century of research, that the differences we see in a population when measuring intelligence, by any means, is around 50 percent. This does not mean that 50

---

* Not quite true. There are many genetic disorders that mean sex is not as binary as this statement decrees: women with Turner syndrome have only a single X and nothing else. Klinefelter's is a syndrome in which males are XXY. Biology is never simple.

† The title of Matt Ridley's excellent 2003 book on the subject. It was originally coined by the behavioral geneticist David Lykken.

percent of *your* intelligence is down to genes, and 50 percent has been determined by your parenting and schooling. It means that half of the differences we see in a population is genetic, and half is environmental.* The point is this: When we looked across huge populations for the genetic component using the GWAS technique, we could only find a tiny bit. The same applied for so many of the complex diseases that we thought, via twin studies (and other techniques), would be sitting in our genes, and the natural variation we were beginning to unravel in the post-genome era. Even relatively straightforward things like height proved to be enigmatic.

We know that height is highly heritable. Dozens of studies over a century have shown that the vast majority of differences in height in a given population are down to genes, and only a small proportion due to the environment. If the tallest person in a population is seven foot, and the shortest five, there's a twenty-four-inch difference. Approximately twenty-two of those inches are the result of genetic differences. Tens of thousands of people have been inputted into GWAS analysis, which has revealed plenty of alleles that stand out as relevant. But, in total, when you add up the contribution that each seems to make to height, it only comes to around an inch or two. The genetic contribution that makes up the other twenty inches is—for now—missing in action.

It's the same for heart disease. And schizophrenia. And drug addiction. And Alzheimer's. And autism. And diabetes. And bipolar disorder. And intelligence. And pretty much any condition, disease, and behavior we have looked at. Genes do not determine the

---

* There is a whole book, probably many, to be written on the heritability of intelligence, and the problems of measuring it, and the possible policies and controversies that follow from this branch of science. This is not that book. I did, however, write and present a series of documentaries for BBC Radio 4 covering this topic, *Intelligence: Born Smart, Born Equal, Born Different*, which are available for free for an indefinite period.

outcome of almost all human biology and psychology. Dozens or hundreds of genes can be involved, each with small cumulative effects, and all mitigated by the world in which we live.

Part of the problem is that we try very hard to categorize diseases together to define them, and to treat them. However, each disease is a unique interplay between the cause(s) and the patient. Some are less unique than others, but ones that involve behavior and psychology and psychiatry are notoriously unique. This is why we reclassify and refine disease definitions all the time. Autism is more correctly now termed autism spectrum disorder, as there is a continuous suite of characteristics that people with autism display. Some are more severe than others; some are merely different.

Some of the missing heritability will be due to limitations in the GWAS approach. It only looks at common SNPs, the points of difference that we see often between people. It might be that much of the missing heritability will be in rare variants. It might be that repeating segments of DNA will account for some of the genetic causes of complex traits and diseases. These so-called copy number variations (CNVs) are backbreaking to sequence, as they are by their very nature the same as a lot of other bits in the genome. Frequently mutations in CNVs are new to an individual and not present in a family history, which also makes them tricky to find.

We don't yet know where the missing heritability is. But it's a scientific mystery, not a supernatural one, and it will be somewhere in the genome; quite possibly we will account for some of the missing heritability in variants we haven't seen yet, in huge numbers with very small effects, that build along with the complexities of a lived life into a measurable condition. Ask me again in ten years' time.

All of these tales play into the same narrative, one that genetics persistently undermines, but culturally we just can't shake. We were so desperate to find simple rules to explain ourselves. The idea was that a gene would encode a trait, or a disease, and we could track these through families, through the generations, and through history. It's an idea that predates genetics by centuries.

The first piece of genetic counseling recorded occurs in the central book of Judaism, the Talmud, in the Tractate Yebamot:

> If she circumcised her first child and he died, and a second one who also died, she must not circumcise her third child; so Rabbi. R. Simeon b. Gamaliel, however, said: She circumcises the third, but must not circumcise the fourth child. But, surely, the reverse was taught; now which of these is the latter? — Come and hear what R. Hiyya b. Abba stated in the name of R. Johanan: It once happened with four sisters at Sepphoris that when the first had circumcised her child he died; when the second [circumcised her child] he also died, and when the third [circumcised her child] he also died. The fourth came before R. Simeon b. Gamaliel who told her, "You must not circumcise [the child]". But is it not possible that if the third sister had come he would also have told her the same! — If so, what could have been the purpose of the evidence of R. Hiyya b. Abba? [No]. It is possible that he meant to teach us the following: That sisters also establish a presumption!

We think that this is a description of what we now know to be the condition hemophilia, where boys are unable to form blood clots easily, and are thus at extreme risk of bleeding to death. It was one of the first diseases to be understood in terms of its genetic cause, in 1989, a condition caused by a single gene on the X chromosome (thus only affecting men; women carrying a defective allele are compensated by their second X). Rabbis, though, recognized it from as early as 200 BCE, saw how the potentially lethal genetic variation was passed on, and gave special dispensation to boys born in hemophilia-carrying families to avoid the procedure. In 1803, a Philadelphian doctor named John Otto accurately described it in a family where the males were affected—he called them "bleeders"—and the females were healthy carriers. Using family histories, he traced it to a single female carrier

> by name of Smith, settled in the vicinity of Plymouth, New Hampshire, and transmitted the following idiosyncrasy to her descendants.

We've seen and known for centuries how traits are passed from parent to child, before we knew of Mendel and his peas, and the rules of inheritance that he would carve out of his leviathan pea-breeding experiments. We knew of the Hapsburg Lip, in fact a marker of the profound and ignoble inbreeding they indulged in, but seen as a signal of royal descent. The first Queen Elizabeth who had red hair, enhanced it further with an array of ginger wigs, possibly after a bout of smallpox had rendered her natural auburn thinning. Some historians believe that she maintained this look to strengthen her visible bond with her much revered and redheaded father, Henry VIII.

Genetics is the study of families, and we have looked into the similarities within generations of families for thousands of years. For most of that time, we've seen and studied the most superficial and the most visible aspects of inheritance. It is not that these traits or these patterns of inheritance are untrue, it is simply that they are the more simple ones.

However, humans are complex: So why would anyone think that human genetics would be simple? Physics has a natural tendency to reduce the universe into simple models that describe everything. We have the Standard Model—a bunch of equations that describe the twelve fundamental particles that all matter is made of, quarks, electrons, and the all-important Higgs boson. It will be refined in time, as new discoveries are made, but most physicists think, based on previous results, that the model of the universe will become simpler again. Biologists sometimes get physics envy, because every time we find out one of our big rules—universal genetics, evolution by natural selection—things look more complex within them as soon as we begin to look.

Mendel found his rules of inheritance in peas, prefiguring the idea of a gene—a discrete unit of inheritance. The Italian experimentalist Theodor Boveri noticed that monstrous forms were created in sea urchins plucked from the bay of Naples if they had an anomaly of chromosomes. In the first decades of the twentieth century, Thomas Hunt Morgan saw in New York that the units of inheritance that coded for eye color in fruit flies sat on these chromosomes, some on the sex chromosomes, some on others. He began to see the complexities of genetics when he saw that genes near each other might well be inherited together. This idea is called linkage (see page 241), and the units carry his name—the centimorgan. Refining that idea from chunks of DNA to individual letters of the genetic code is one of the ways we see evolution in action, in us and in our ancestors.

In the 1990s, genes were discovered that told us how to put together a human or a fly or a worm. They were remarkably similar in all these beasts, and seemed to have the same function wherever we looked; a head at this end, a tail at the other; this is going to be an eye; this is your front, this is your back. Even in the difference between insects and mammals, we could see the same genes doing the same thing. These beautiful genes were incredible indicators of, not human history, but the history of life on Earth. They show evolution as a tinkerer, in the lovely phrase of the geneticist François Jacob. When something works in one way, it might be co-opted to work in another. The Hox genes, and the Pax genes and the master control genes were revolutionary in terms of how we understood shared ancestry. How bizarre that a gene could simply say "this tissue is going to be an eye" regardless of what type of eye that would be. How unlikely that the genes that say heads and tails were the same regardless of whether your head looks like mine or a worm's.

Even this wonderful branch of discovery reinforces the notion that there are genes for things, for heads, or eyes, or tails. All of the genetic programs set up by these master controllers are entirely context dependent. They may have definitional power, but a fruit fly with a mouse PAX6 inserted in it grows a fruit fly eye, not a mouse eye. Every step in the history of genetics reinforced the idea that there were genes for everything we could imagine or observe. And this pushes the idea that the presence of these genes for things meant destiny: Your fate is in your genes.

Well, that's not what genetics now says. This deceptive narrative continues to this day, and into the current era of consumer genetics. When I had my genome explored by 23andMe, it revealed some oddly specific things. I don't flush or puke when I drink alcohol (in small measures). I have wet earwax. I don't have alleles for cystic fibrosis, Tay-Sachs, sickle cell, or Gaucher's disease. I have a 28 percent chance of having blond hair. What that actually means is that 28 percent of people with exactly the same allele at that exact same point in our genomes have blond hair.

One thing slightly stood out: I am not a carrier of an allele that causes a lung disease that appears in my family by marriage, which is of some relief as it greatly reduces the chances of my children suffering from or passing on that disease.

Most of this stuff vaguely contributes to the cultural half-truths of the way we talk when we talk about genetics. I have brown eyes, which is a trait dominant over blue. When I learned patterns of inheritance at school, we were taught how blue eyes are a recessive allele, and brown is the dominant version. Therefore, if you have one of each, you would have brown eyes. If you had two blues you have blue eyes. And if you have you two browns, then you have brown eyes. All very straightforward, and correct.

We draw these possible outcomes on a neat table, a Punnett square, devised at the beginning of the twentieth century by Reginald Punnett. It allows us to work out the wrinkliness of a pea, or the chances of passing on cystic fibrosis, or the likelihood of the eye colors of the children of parents with various combinations of eye color genes. Two brown-eyed parents can still produce a blue, but only if they are both carriers of a blue allele and, even then, the chances are only one in four. Three out of four possible outcomes will result in a brown-eyed girl or boy.

We were also taught that the ability to roll one's tongue showed similar patterns of inheritance. Other schools taught that the bend in the distal joint of the fifth digit was another—the so-called hitchhiker's thumb—or the attachment of earlobes to the skull, the dimple of the chin. Clasp your hands together with interlocked fingers; one thumb will be on top of the other, and if you try to reverse them, it'll feel impossibly weird. This has also been taught over the years as a trait inherited via a single gene.

Mendel set up these proportions in his leviathan pea-breeding experiments in Brno in the 1860s. The rules he came up with are correct. Characteristics are encoded in genes in discrete units, and they are inherited discretely. The combination of two different alleles of the same gene that result in different phenotypes does not result in a blend of those two, but in a likelihood of one emerging over the other according to Mendel's laws—pea wrinkliness, flower color, tongue rolling, eye color. The trouble is that we are more complex than peas. And these traits we are taught are not necessarily so clear-cut. For tongue rolling, some people can, and others can't. But we've known that this trait is not a simple Mendelian characteristic for decades, since studies in the 1950s with identical twins. Out of thirty-three pairs, seven were discordant for tongue rolling—identical siblings, one of whom could and the other couldn't perform this tremendously important dark art. Alfred Sturtevant, one of the giants of genetics in the first half of the twentieth century, had first suggested tongue rolling as a Mendelian trait in 1940, a single allele bestowing the ability on its bearer or not. After the twin studies he, like a good scientist should, changed his mind, and said in 1965 that he was "embarrassed to see it listed in some current works as an established Mendelian case." It is still taught in schools today.

Twin studies also showed that identical siblings vary in the way they clasp their hands. Ear lobes are not either attached or not: Some are, some aren't, most people are in between. Elements of the genetics of eye color are true. Brown is indeed dominant over blue, blue is a recessive condition and two blue-eyed parents cannot produce a brown-eyed child. But there's also another gene that encodes greeny-hazel eyes. What this means is an almost complete spectrum of eye color from the palest blue to the darkest brown, and though much of the variation in our eyes can be attributed to a relatively simple pattern of inheritance, it is not a straightforward Mendelian trait. So far, ten other alleles have been shown to have an effect on eye color, which means that making predictions on what color a child's eyes will be based on their parents' is not something I would bet on. When it comes to humans, the simple, direct, understandable answers are likely to be wrong.

This science needs you

In 2001, some journalists characterized the revelations of the first round of results from the HGP as some kind of a calamity in human genetics. The betting book was not in the public realm, and no one really knew about it outside the world of genetics until years later. Nevertheless, some journalists chose to report the HGP as the discovery that everything we thought we knew about genetics had turned out to be wrong, and it was a field in the grip of a catastrophe. Others since have declared it a waste of money, because we had discovered things that we didn't anticipate. The same happened when the GWAS studies began to show a picture of complexity that we hadn't foreseen. A writer in the *Guardian* newspaper in 2011 said:

> Among all the genetic findings for common illnesses, such as heart disease, cancer, and mental illnesses, only a handful are of genuine significance for human health. Faulty genes rarely cause, or even mildly predispose us, to disease, and as a consequence the science of human genetics is in deep crisis.

There is no crisis. There is only science. A popular celebrity therapist in the UK named Oliver James repeatedly asserted in the UK national press in 2016 that genetics played little or no role in the inheritance of behavioral traits and psychological disorders. He was gifted many platforms on TV, radio, and in print to state this as fact, by people who were hungry for this controversial and heterodox view. He is wrong, very easily demonstrably wrong. He and others who repeat this misunderstand what the problem of the missing heritability is, and have fallen into the tempting trap of mistaking absence of evidence for evidence of absence. We swing from genetic determinism to genetic denialism.

Both extremes are simplistically wrong. It is true that we can't yet account for the predicted amount of "nature" for a whole suite of traits. This is science, and what scientists do is work out what they know and what they don't know, and slowly move things from the latter category into the former. We are all Rumsfeldian, if we are doing science right.

No single technique was ever going to explain a human, and that quest is endless. Geneticists are dealing with the longest, most complicated coded message ever written, about the most complicated entity ever to have existed. The comedian Dara O'Briain has noted several times in response to the fatuous attack "scientists don't know everything" that it is totally true: If we did, we'd stop. Each technique has merits and weaknesses. All studies are flawed in some ways, and all results in science are always conditional. Studying the genome has been a wonderful exercise in humility. There are no eureka moments, there are few truly revolutionary findings, and paradigms shift—to borrow that sticky meme from the sociologist of science Thomas Kuhn—at glacial rates. We chip away at the edges of the unknown, a never-ending jigsaw puzzle—science as a way of knowing.

We are culturally programmed to misunderstand genetics, from the Hapsburg Lip to the way we talk about inheritance. For the most part it matters for science. We should care about learning and understanding things so that they are as up to date as possible, and not beset by myth or misunderstanding. It matters too because as access to our genomes becomes quicker and cheaper, we are presented with raw data—risk factors for diseases and characteristics. It's easier to fork out £100 for this than it is to learn the complexities of genetic pathways, and epigenetic interactions with the environment, and the statistics that underlie risk and population dynamics. We perpetuate myths by clinging to these simple stories, and fail to bathe in the wondrous complexity of what it means to be a human.

We do at least now know much of what we don't know, like an alcoholic admitting that they have a problem. Two decades ago, we didn't, as Ewan Birney's betting book revealed. That is progress. And there is so much to find out.

If you want answers, the best thing you can do is *join us.* The technology that was invented during the HGP and beyond has generated more genomic data than we can currently process. Huge arrays of DNA sequence are just sitting there waiting to be mined. Much more is waiting to be read, nowadays from the living and the long dead. And it's only going to increase. I believe that the way forward in human genetics is to sequence everyone in full, from birth. We would then enter a realm of data security and privacy issues that are yet to be fully realized, and these need to be tackled in society with the widest possible consultation. But we are effectively infinite in our variety, and to pursue that difference, in characteristics and disease, in inheritance and history, we need all the data we can get. And then we need to crunch it.

Humankind is a species of explorers. For millennia, we've built and invented and tested in order to try to understand the world, the universe, us and our place in it. There are plenty of frontiers left to explore, but let's face it: You're almost certainly not going to be an astronaut, and neither are your children. Space is hard, and the selection criteria so tight that fewer than a thousand people in human history have made the grade. But there's a frontier inside every cell, and you can be part of that exploration. The only completed map is one of a dead planet, frozen in time. Our genetic map is changing continuously, with bodies dug up from the recent and ancient past, with continuing migration and the admixture of people of the world, and with the never-ending fight against cancers and other diseases that will, one day, be merely of interest to historians. We need geneticists and mathematicians and computer scientists and coders, archaeologists, doctors, and patients. Come with us.

7
Fate

"Yet man is born unto trouble, as the sparks fly upward."
Job 5:7

October 16, 2006, Kimsey Mountain, Polk County, Tennessee

Davis Bradley Waldroup waited at his trailer home in the mountains of Tennessee for his estranged wife Penny to arrive home with their four kids. Their relationship had been strained for months, and he had been drinking heavily. When they arrived, with her friend, Leslie Bradshaw, he appeared carrying a .22 rifle and they began to fight. Penny said she was leaving. He removed the keys to Penny's van and threw them into the woods. Minutes later, Waldroup shot Bradshaw eight times with a rifle. She died. Penny attempted to escape up the mountain, but he shot her in the back as she ran and cut her with a pocketknife when he caught up with her. He then bludgeoned her with a shovel, and then a machete, slicing her dozens of times, and chopping off one of her fingers.

Waldroup dragged her to the trailer wanting sex, and became angry that she was unresponsive and too bloody. He told his children, "Come tell your mama goodbye." But she managed to escape.

The police arrived a few minutes later to a scene of carnage: blood on the walls, the truck, the carpets, and even the Bible that Waldroup had been reading while drinking that evening. On top of the gruesome physical evidence, of which there was plenty, he had expressed clear murderous intent, and in Tennessee, he faced execution.

March 25, 2009: After eleven hours of deliberation, the Polk County Grand Jury handed down a verdict of aggravated kidnapping and voluntary manslaughter of Leslie Bradshaw, and aggravated kidnapping and attempted second-degree murder of Penny Waldroup. Bradley Waldroup had escaped conviction for first-degree murder, and had avoided execution. He still faced thirty-two years in prison, and the judge advised against an appeal on the grounds that the jury might not be so forgiving next time. He was subtly warning that the complexities of the case were on arguable foundations. The defendant's experts had claimed that he could not have committed first-degree murder because he was "unable to engage in the reflection and judgment necessary to premeditate the crimes." The reason why, according to Waldroup's legal team, was genetics.

At the root of their defense was monoamine oxidase A, known as MAOA. The gene MAOA encodes an enzyme whose role is to destroy molecules called neurotransmitters. These are the messengers that flit between synapses in the brain when we think or do, and regulating their activity is an essential part of a normal life. Serotonin, dopamine, and noradrenaline are the best known neurotransmitters, though there are more than a hundred, and these are some of the molecules that are involved in making us happy or sad, or affected by drug use, or sex, or diseases. MAOA simply clips part of these molecules off once the message they bear has pinged between cells, and in doing so renders them done. It's like shredding a letter once you've received and read it. Typical for the molecular tools that are continuously employed at this level of brain activity, MAOA is essential for a normal life. When it's not working at full capacity, or when unusual genetic variants are at play in a person's neurons, all manner of problems can transpire. It's associated with cardiovascular disease in some patients, with cancer in others. When it comes to the way we act, MAOA has been linked to autism, Alzheimer's, bipolar disorder, attention deficit hyperactivity disorder, and serious depression. This kind of plurality is not unusual in the molecules of the mind, but it does show how complex and sometimes inscrutable the relationship between genes and behavior is.

Toward the end of the twentieth century, reports began seeping out of labs that particular versions of MAOA were turning up more often in people with aggressive, impulsive, or criminal behavior. It all started in Nijmegen with the men in a large nefarious Dutch family. Five generations of this dynasty, all the way back to 1870, were packed with criminals, including arsonists, attempted murderers, and a rapist. IQs of the men in this family were frequently low, typically around eighty-five, which is well below population average and borderline for intellectual disability. Women in the family were expressing fear that the threatening behavior of some of the men was endangering them. Han Brunner, a doctor in Nijmegen, examined the family histories, and narrowed down a shared anomaly on the X chromosome. This went some way to explain why the men were afflicted and women not—many conditions that affect men and not women are called X-linked, as the root cause lies on the X chromosome of which men have only one copy; a woman's spare X can sometimes compensate if they bear the genetic anomaly on the other X. The zone on the X, at least in the pre-genome era, contained many genes, but MAOA stood out as a candidate. Its role in digesting neurotransmitters made a robust case for it being involved in behavior, and this was known from earlier studies in mice. When Brunner tested the urine of the Dutch men, he found significantly low levels of the normal molecular leftovers following the action of MAOA—not enough shredded letters. It appeared that it was not working as it should.

Over the years that have followed, the resolution of the genome became clearer and the studies of MAOA snowballed. In 2000, the association was strengthened with a study showing it was not the coding of the gene itself, but in its promoter. That's the chunk of DNA before the beginning of the gene that receives the cue instruction to turn MAOA on. For MAOA, the promoter contains a short repeating section of DNA, and men who scored highly on aggression and compulsive behavior questionnaires tended to have fewer of these repeats in their MAOA promoters.

MAOA acquired the nickname the "warrior gene" around 2004. Blaming the tabloids for bastardizing or distorting science is frequently not unreasonable, as they are prone to magicking away nuance, or conjuring up meaning. But in this case it was the august academic journal *Science* that instigated the frothing media frenzy that would characterize the debates about genes and crime for a decade to come. A study in monkeys had uncovered variants of MAOA, and postulated that the aggression that fell out of a defective gene might be an advantage in battling other monkey troops. There, in the headline was the label: "warrior gene."

It really took hold. In 2007, a small study in New Zealand claimed that a particular version of MAOA was common in Maori men, the indigenous Polynesians who populated the islands from the thirteenth century onward. The researchers stated that:

> It is well recognized that historically Maori were fearless warriors.

"It is well recognized" is one of those phrases that make me a bit twitchy. When I read that in a scientific journal, in my head I translate it to "we couldn't find a reference for this": It might be true, but it's difficult to substantiate. The study comprised just forty-six men, who needed to have only one Maori parent to be classified as Maori. Seventeen of them were isolated who had eight Maori great-grandparents as a means of refining the Maori DNA, and this group was tested for the MAOA allele to establish the sweeping generalization that has become a cultural meme: Maori are genetically predisposed to violence.

The Maori do indeed have a reputation as a historical warrior culture, and when you see the terrifying war dance haka at the beginning of an international rugby match, big men quake. But "fearless warriors" doesn't sound like a scientific statement to me. It sounds like a preconceived value judgment, and impossible to quantify. Though New Zealand is sometimes held up as one of the better examples of European colonizers respecting the traditions of indigenous people, prejudice against Maori in New Zealand is very real, and reinforcing stereotypes in this manner is perhaps questionable in a scientific paper. The ethnic profiling against MAOA extended beyond the Maori and showed that the Chinese had the highest rates of defective MAOA—77 percent—and white Caucasians the lowest at 34 percent. Do the Chinese have a reputation as being genetic warriors?

A chunky paper in 2003 suggested that violence associated with the defective variant was significantly worsened if the perpetrator had been sexually abused as a child, whereas abused children with a normal MAOA were less likely to be criminals. This result was affirmed in 2012 in a meta-analysis—a super-powerful way of aggregating multiple studies to ramp up the analytical depth.

And the studies keep coming, with subtly specific conditions, with the law trailing behind them. In 2009, an Algerian living in Italy named Abdelmalek Bayout had his sentence for the murder of a Colombian man reduced by three years, after his defense also identified him as a carrier of the defective MAOA gene. An experimental study in the same year used the "hot sauce paradigm," a common technique in experimental psychology where spicy chili is used as a proxy for punishing potential enemies. Hot sauce has been deployed in all manner of hypothetical conflict over the years, carefully meted out as punishment, by researchers unable to actually hurt people experimentally. They were testing the impulsivity and willingness to punish people who had supposedly taken money from them. Those with the defective version of MAOA handed out more chili quicker.

An American study in the same year showed that gangs in the United States showed high proportions of members with defective MAOA, and, of those, they were more likely to use weapons in a fight. They also showed that 40 percent of gang members had a normal MAOA gene, and most defective MAOA carriers were not in gangs. They concluded, admittedly after mentioning some of the limitations of this preliminary study, that their

> investigation shows that gang formation and activity, like most antisocial behaviors, involves gene-environment interplay.

One could satisfactorily replace that phrase *like most antisocial behaviors* with the overarching qualifier *like all human behavior*. I find it very odd that a scientist would make that distinction.

Even the august journals are tempted into these grotesque simplifications. In 2008, *Nature* reported the discovery of the "ruthlessness gene," accompanied by pictures of Hitler, Robert Mugabe, Saddam Hussein, and Mussolini. In 2010, another serious academic journal (*EMBO Reports*) dismissed the idea of a "criminal gene" in its first sentence, though the title mutated warrior genetics to "The Psycho Gene." To its credit, it also presented a nuanced but robust view of the relationship between genetics, behavior, and the law:

> "Taking genetic factors into account when sentencing is plain stupid, unless we are talking about something like Down's syndrome or some other syndrome that drastically reduces intelligence and executive functioning," insisted Anthony Walsh from the Criminal Justice Department at Boise State University in Idaho, USA. "This is the kind of 'genetic determinism' that liberals have worried themselves silly over. They just have to take one or two neuroscience and genetic classes to dispense with their 'my genes/neurons' made me do it. Nothing relieves one of the obligation to behave civilized."

It's not just criminal violence or impulsivity. MAOA has been the subject of a clutch of studies beyond aggression over the last few years. A minuscule association between a variant of MAOA and addictive gamblers was published in 2000. In 2009, there was an association with long-shot risk taking. Without any further mechanistic analysis, the assumption proposed was that this allele predisposed people to risk-taking, impulsive behavior. All these studies are flawed, as are all scientific papers to some degree. Many of them are weak statistically, or underpowered, or simply have absurdly small sample sizes. That may not make them wrong, but it doesn't help them to be right.

In Finland in 2014, 800 violent offenders, their crimes amounting to 1,154 murders, manslaughters, attempted murders, or assaults, were screened by a team lead by Jari Tiihonen from the Karolinska Institutet in Sweden. Again, MAOA glimmered from the dark of the genome, as well as an allele of a second gene, called CDH13. The study had used a GWAS approach to hunt for genetic clues to violent crime, and these two genetic variations, and the Manhattan plots, are often apparently chaotic, with the notable peaks only just nudging into the sky past the melee of statistical insignificance.

CDH13 is a gene whose protein is involved with neurons making connections in the brain, notably in the amygdala—two symmetrical grape-sized parts of the brain known to be involved in emotional responses, decision-making, and memory. Defects in the amygdala have been associated with fear, aggression, and alcoholism, all of which might appear to line up conveniently with violent criminality. Other amygdala defects associate also with anxiety, sexual orientation, and violation of personal space. CDH13 is also thought to be one of the key players in attention deficit hyperactivity disorder (ADHD), which is characterized by impulsive behavior. Of the Finland cohort, 80 percent of the murders were not premeditated, but were impulsive. Does this begin to fit into a bigger picture? Not really. These are at best speculative associations in a gargantuan jigsaw. Some might be relevant, others might not be. They might be factors, some even might be causes, but for now, we don't know. The authors of this study are very lucid on this point, robustly stating that the associations they find are statistically way too low to be considered for any conceivable screening purposes for identifying potential violent criminals. And they even go further:

> According to the basic principles of forensic psychiatry, only the actual mental ability (phenotype) of the offender matters when punishment or legal responsibility is considered, and putative risk factors *per se* (such as genotype) have no legal role in the resulting judgment.

Jari Tiihonen told the BBC in 2014, "Committing a severe, violent crime is extremely rare in the general population. So even though the relative risk would be increased, the absolute risk is very low."

All these studies point to possible factors in the etiology of violent crime, that may under the right (or more precisely, wrong) circumstances agglomerate into a tinderbox combination. Behavior is complex. Genetics is complex. Real world narratives mostly fail to recognize that cumulative complexity. Bear in mind that according to one of the studies, one third of white men carry the same allele as those who murder or fight, or dish out hot chili sauce as punishment. Statistically speaking, none of them will murder. Bradley Waldroup carries that version, the so-called "warrior gene." He was beaten and abused as a child, horrifically. He had been drinking heavily on that night when he killed and maimed. He had ready access to firearms and other weapons. I do not support capital punishment. But I have no doubt that the route by which he and Abdelmalek Bayout were at least partially exonerated from heinous crimes was wrong. One juror in Waldroup's trial was quoted as having said:

> A diagnosis is a diagnosis, a bad gene is a bad gene.

Despite all the studies, we simply do not know well enough how this gene works, how it participates in the biological melee of a life, how life experiences and chance coordinate with the external world of people. Even if we did, the legal ramifications would be equivocal, and subject to political leaning. Are we slaves or masters of our genes? We are neither, and it's a dumb, simplistic question. To say otherwise is a biological determinism with profound legal consequences. Of course, this is not new to the justice system, it is merely that now genetics is coming of age, it is now part of the pantheon of legal defenses and excuses for predestination. A gene that has a measurable association with violence if the bearer was beaten as a child is not irrelevant. But perhaps exoneration via the complex and poorly understood root of genetics is missing the broader point that maybe we shouldn't abuse children.

On December 14, 2012, in what is now a commonplace occurrence in the USA, twenty-year-old Adam Lanza loaded handguns, a shotgun, and semiautomatic assault rifles and drove to a local school in his hometown of Sandy Hook, Connecticut. There he murdered twenty-six people, including twenty primary-school-aged children, and finally shot himself dead with a pistol. He had already murdered his mother in her bed before he drove to the school. Immediately, in the wake of this almost inconceivable crime, speculation and analysis began about his motivations, and the causes of such a wicked act. Video games are often the first port of call for attributing blame in spree killings. By December 20, a US senator had introduced a bill calling for research into the relationship between video games and violence* on the grounds that "Parents, pediatricians, and psychologists know better."

Lanza was indeed a gamer, and earlier on his final day had been playing Call of Duty 4, a very violent military shoot-'em-up game, as he frequently did. The press clamored to provide some exclusive insight and explicatory profile to a public desperate to understand or apportion blame. It's a strange argument because Call of Duty had at that point sold 15 million copies worldwide,† and indeed the penetrance of video gaming for teenage boys in the United States was around 84 percent in 2015—meaning that the vast majority of teenage boys play video games. Even with the soul-crushing frequency of spree shootings reported, the statistical relationship between playing violent video games and enacting murder is

* This is an area of research that is characterized as being "opinion rich, data poor." A paper, which I coauthored in 2016, attempted to address the question of levels of violence in video games and subsequent delinquent behavior in teens. We found a very weak effect, which was probably underpowered. It's not a very headline-grabbing result, but that's science, folks.

† Indeed, I have played earlier versions in the frequently excellent Call of Duty series for many, many hours.

irrelevant. Because games are so widespread, statistically, the idea of a causal link between violent video games and murderous violence is absurd. According to the final report into the shootings, Lanza also enjoyed playing Dance Dance Revolution and Super Mario Bros., the former a game in which you have to mimic disco moves, and the latter one in which you have to collect gold coins and rescue a princess.

In the quest to source blame, the local authorities next turned to genetics. Not to MAOA or an alternative single gene, but, now that whole genome sequencing is commonplace and cheap, all of his DNA. It was reported in the press that Lanza's genome was to be sequenced to find out if his evil lay in his DNA. Was he born to murder?

Adam Lanza was a profoundly disturbed boy. He was diagnosed with sensory processing disorder, a condition in which children have significant sets of difficulties in engaging in normal social behavior, the autism-spectrum disorder Asperger's syndrome, and obsessive-compulsive disorder. All of these conditions have genetic elements to them; none of them is caused singularly by genes. All studies so far have shown that many genes of small effect play a very small role in the diagnosis of any of them. That is a very typical statement when it comes to complex psychiatric disorders. There is no gene for any one of them, and any association found is weak. Furthermore, there is no known association between any one of those disorders and violent crime.

Nevertheless, *The New York Times* reported on Christmas Eve 2012 that researchers at the University of Connecticut would be sequencing Lanza's genome. Details were scant, though a spokeswoman confirmed their plans, and, to date, the results have not been made public. Response from many scientists was abrupt and dismissive. Robert Green, a geneticist and neurologist at Harvard Medical School told *The New York Times* that for mass murder:

> It is almost inconceivable that there is a common genetic factor. I think it says more about us that we wish there was something like this. We wish there was an explanation.

The writer and scientist P.Z. Myers added a succinct preemptive conclusion, one with which I wholeheartedly agree:

> I can predict exactly what will be found when they look at Adam Lanza's DNA. It will be human. There will be tens of thousands of little nucleotide variations from reference standards scattered throughout the genome, because all of us carry these kinds of differences. The scientists will have no idea what 99% of the differences do.

This exercise is futile. It is pointless bluster dressed up as science. We could find out his probable hair color, or the nature of his earwax, or a handful or other traits. We might be able to find out his chances of developing a handful of diseases. But even if there were a thousand or ten thousand shootings near identical to Lanza's crimes, and all of those genomes were deciphered, and they turned out to have exactly the same statistically significant differences from people who don't commit spree killing rampages, then, based on what we know about much simpler human traits, the genetic aberrations would still probably account for a tiny proportion of the heritability we might see.

Even if this were the case, and I would bet my house that it couldn't be, then it is fully unclear to me what necessary policy implications this would suggest. Would we then screen for this precise genotype in newborns? If a person had 99.999 percent of the same genotype as Lanza, would we put them on a watch list? Waldroup and Bayout had their sentences ameliorated in light of their genes. In a different political climate, would it not be equally justifiable to increase their sentences on the grounds that they are simply born bad? Would it not therefore be arguable that they could never be cured, and that they therefore should be sterilized or executed, as so many were in the USA and Nazi Germany under their vigorous embrace of eugenics as policy? This facile trend relies on a simplistic misunderstanding of genetics. It is a probabilistic science, and any action predicated solely on DNA sequence seems to me to be a high-risk endeavor, prone to specious failure.

We look to statistics for reassurance in these types of case. Here is one: 100 percent of mass shootings have been enabled by access to guns. I can guarantee that even if there were a genotype shared by the mass shooters, which there will not be, none of the killings would have happened if they didn't have guns.

These stories are only as new as the science itself. We have been trying to pin criminality and all manner of complex human behaviors on biology for centuries. These days it's genetics and often brain scans being presented as some kind of definitive proof of the root cause of a particular behavior, such as religiosity, or envy, or love. Those pretty scans of a slice through a brain we are so familiar with nowadays, with a little dot in a different color "lit up," as has become the standard press phrase to describe a spot of neurological activity that supposedly corresponds with whatever the owner of that brain is being asked to do or think about. These types of scans, magnetic resonance imaging (or when a specific task is being asked of a subject, functional MRI) are, like genetics, mostly robust good science and well understood within the domains of science. It is not as simple as just sticking someone in a scanner and asking them a bunch of questions, and seeing how the pictures come out. Just as we can't simply read genes and infer how someone will behave, we can't read minds. The scans read oxygen consumption in the brain, as a proxy for which bits are working hard. But just like genetics, they require careful statistical analysis and correction before any conclusions can be made. The subtleties of brain activity, and the resolution of thought at the cellular level are precise and minuscule. These brain scans are informative but not definitive, though this is not the impression one might get from reading the news. Placing them in the context of court, where judgments rely on evidence, the doubt so crucial to science confounds decisions.

John Wayne Gacy was one of the most prolific US serial killers. Based in and around Chicago, he preyed upon teenage boys and young men, violently and sexually assaulting them. After he was executed for the murder of thirty-three boys in the 1970s, his brain was removed to be studied to see if his evil lay in the structures of his gray and white matter. Before him, in 1929 in the Rhineland in western Germany, Peter Kürten murdered at least nine people, and assaulted many others, in a way that earned him the legend the "Vampire of Düsseldorf." He was beheaded by guillotine in 1931, and his brain was removed too, to see what evil lay within. Today, his mummified head is on display in the Ripley's Believe It or Not! museum in Wisconsin. Neither brain yielded any clues to their crimes. They were human brains.

The desire to understand extraordinary human behavior via genes is not just limited to violent murderers. In 2014, the British television station Channel 4 broadcast a prime-time series called *Dead Famous DNA*. Major figures from history were selected as quarries, and the hunt, with suitable portentous music, for verifiable pieces of tissue from Beethoven, John Lennon, Napoleon, and Adolf Hitler was the meat of the show. The objective was to suck out their DNA to see if it offered clues to their globally significant achievements.

What a strange eldritch world is the business of dead celebrity body parts. *Dead Famous DNA* was ghoulishly entertaining, when not troublingly amoral. The overwhelming sense for the viewer was of amusing schadenfreude, mostly at the expense of the program itself. Television shows often rely on a metaphorical journey of discovery, but this was a televisual journey of discovering expensive folly. Not least when Napoleon's severed penis was deemed too small for useful DNA extraction—they would've needed more than an inch of this shriveled flesh. Nor when the $10,000 spent on Marilyn Monroe's and John F. Kennedy's hairs was blown, as they had been sun-bleached free of DNA. Or the $5,000 sample of King George III's hair that turned out to be from a wig. But it was distasteful and morally repugnant to buy hair from a grotesque prominent Holocaust-denier and Nazi regalia trader, though viewer disgust was slightly mitigated when it turned out not to be from the head of the Führer, but from an Indian man.

Tracing the flesh of the dead wasn't the main point of this genetically illiterate stunt though. "Could their DNA reveal what made Marilyn Monroe so attractive, Albert Einstein so intelligent, or Adolf Hitler so evil?" the furrow-browed presenter intoned with concerned gravitas.

"No," is the answer. Admittedly that would've been a shorter program though. Is evil encoded in DNA? No. Is intelligence? It certainly has a heritable component, a significant one, but that is measured across populations, and not in individuals, and the hunt for specific genetic correlates of intelligence has not yet been particularly fruitful. As for beauty, well I don't fancy Marilyn Monroe, as, I believe, many people don't around the world. I'm more of a Lauren Bacall man. I seem to recall that there might be a phrase, something to do with "eyes" and "beholders" that encapsulates this idea. There are indeed archetypes of female beauty, but many of these are culturally specific, and far from universal even within those cultural bounds. Marilyn's physicality was partly in her genes, and that physicality is partly her allure, no doubt. But her profile, acting, personality, clothes, makeup, hair, and all the other things that contribute to her perceived beauty were not encoded in her DNA, nor could be revealed with genomics. Before she became Marilyn, Norma Jean was a redhead, but I'll wager that in arguing for her status as an archetype of female beauty, it is her iconic bottle blonde hair that is under scrutiny.

It doesn't matter whether we are talking about criminality, or psychological characteristics, or psychiatric disorders, or perfectly normal human behaviors like political bent, or susceptibility to alcohol, or being gay or anywhere on the spectrum of sexual preferences, the biology that is revealed by genetics are not causes, or triggers, or foundations. They are potential factors: probabilities.

We hope that science is nuanced and complex, and not prone to trends. That, alas, is not true. Techniques become easier, and adopted widely. The scrutiny applied varies, in labs, in people, in publications, in funding. Journals are not all equal, and publication in a journal is not a mark of truth, merely that the research has passed a standard that warrants entering formal literature and further discussion with other scientists. That also means in public, and, rightly, in a press for whom nuance and scrutiny are not held to the same standards. Genetics matures, and the data flows. It is complex, and unclear, and requires analysis, and parsing, and experimental testing, and more of all of that over and over. That relationship between a complex ecosystem of scientific research and publications and how that information enters the public domain frequently is broken. H. L. Mencken's maxim at the beginning of Chapter 6 captures it perfectly, and ironically given that he was a newspaperman:

> For every complicated problem there is a solution that is simple, direct, understandable, and wrong.

No one will ever find a gene for "evil," or for beauty, or for musical genius, or for scientific genius, because they don't exist. DNA is not destiny. The presence of a particular variant of a particular gene may just have the effect of altering the odds of any particular behavior. More likely, the possession of many slight differences in many genes will have an effect on the likelihood of a particular characteristic, in consort with your environment, which includes all things that are not DNA.

Google "Scientists discover the gene for" and wander through the headlines returned in their thousands, from every media outlet, from the trashy to the august.

> SCIENTISTS DISCOVER THE GENE FOR COCAINE ADDICTION
> *Guardian*, November 11, 2008
>
> SCIENTISTS DISCOVER "TRANSSEXUAL GENE" THAT MAKES MEN FEEL LIKE WOMEN
> *Daily Mail*, October 27, 2008
>
> SCIENTISTS DISCOVER HEIGHT GENE
> *BBC Online*, September 27, 2007
>
> STUDY: A GENE PREDICTS WHAT TIME OF DAY YOU WILL DIE
> *Atlantic*, November 19, 2012
>
> GENE THAT CAN SCARE YOU OUT OF YOUR MIND: SCIENTISTS HAVE DISCOVERED AN "ANXIETY GENE" THAT MAKES PEOPLE MORE FEARFUL
> *Daily Mail*, July 19, 2002
>
> SCIENTISTS DISCOVER HOW KEY GENE MAKES PEOPLE FAT
> *USA Today*, August 19, 2015
>
> SCIENTISTS FIND "GAY GENE" THAT CAN HELP PREDICT YOUR SEXUALITY
> *Daily Mirror*, October 9, 2015

Sexual behavior is a spectrum. At one end some people are exclusively homosexual and at the other end some are exclusively heterosexual. Most people are somewhere in between in thought, word, or deed. Some people are asexual. Sexuality is, like everything else, partly heritable, and difficult to measure. Twin studies indicate the heritability is around 50 percent—roughly meaning that identical twins were both gay about half of the time. Some people consider themselves always to have been gay, sometimes since before they were sexually aware. People's behavior changes over time, and that means that heritability also changes over a lifetime. Polymorphisms have been suggested from studies that associate with homosexual behavior, but again, these are probabilistic genotypes. Being in possession of one of these genetic variants will not make you gay, just as an absence will not make you straight.

These simplistic narratives are just wrong.* They're primarily drawn from GWAS research, described in Chapter 6, and while there is an element to them that is extracted from research, the headlines are often not in any way an accurate reflection of the results of the actual studies. Some diseases do have a single root cause in a single gene, but how that disorder manifests can be highly variable, a concept in genetics called "penetrance." As discussed, even the most genetically straightforward diseases, such as cystic fibrosis, are mitigated by a host of other factors buzzing around inside our genes, cells, and outside our bodies. Inheritance is a game of probability, not of destiny.

It's not just the headline writer's fault though. The history of science is clearly to blame too. Recall Gregor Mendel, who gave us the rules of inheritance by studying individual characteristics in pea plants. Through the twentieth century we beavered away at the

* For no other reason than utter vanity, I have proposed this fallacy might be known as Rutherford's Law, in the style of generalized Internet rules such as Godwin's Law (the tendency for an online discussion to cite Hitler increases as the discussion continues) or Betteridge's Law (if a headline poses a question, the answer is likely to be no). My suggested narcissistic law could be stated thus: if a headline states that "Scientists have discovered the gene for X," where X is a complex human trait, they haven't because that gene does not exist.

laws of inheritance, and unraveled DNA, and cracked the genetic code. In the 1980s, the first disease genes that we identified were indeed "for" specific diseases—cystic fibrosis, Duchenne muscular dystrophy, Huntington's. We tracked eye color through families, and other traits that we see in families and in the mirror. But twenty years after these revelations, in the era of the genome, we discovered that these were the unusual ones. We had characterized them first exactly because they run in families in a straightforward way, because they were easier to spot. Most human traits, behaviors, and diseases are complex, with dozens or hundreds of genes playing a small part in concert with the inscrutable milieu in which they operate.

The simple deterministic version of genetics is merely a new phrenology. In the nineteenth century, largely led by George Combe in the UK and the US, the shapes of people's brains and skulls was promoted as a means of determining human traits, such as cautiousness, self-esteem, truthfulness, or conscientiousness. The trend lasted a few decades, and drifted out of fashion toward the time of Darwin and Galton, who was dissatisfied with its behavioral inaccuracies. In Italy, though, Cesare Lombroso clung to the belief that criminals were *reo nato*, born not made, and that they could be identified by a whole range of physical attributes. The crimes ranged from rape to murder to theft, the evidence included the shape of the brow, the asymmetry of the face or skull, the length of your arms, the shape of your ears.

But his measurements were poor, the fashion passed, and phrenology is now correctly dismissed as a pseudoscience. It has the sheen of fact, but no depth. The simple testing of the ideas of phrenology proved it specious: The measures of difference were insignificant, and the measures of personality whimsical. Phrenology was slain by data. It was an idea, nothing more, that did not stand up to the scrutiny of science. Anatomy went from being macro to cellular as microscopes developed, and then molecular when we began to understand how DNA works. The principles still apply: It is not possible to predict the complex behaviors of someone solely by the bumps and bulges and basic morphology of the skull. And it's not possible to do it with DNA either.

The western Netherlands, November 1944

It was the endgame of the war. The Third Reich was coming to the end of its wicked assault on Europe. The D-Day landings in June had put the Allies into German territory and were forcing the Nazis to retreat. Parts of southern Netherlands had been liberated, but the Allies' plan to end the war by Christmas had failed. Operation Market Garden was their attempt to enter Germany across the Rhine, but at Arnhem and elsewhere they were unsuccessful at capturing the bridges needed to push the enemy back into their own homeland. In September, the Dutch railway workers went on strike, in an attempt to weaken the Nazis' grip on the dwindling occupied territories, but the Nazis had one last bitter retaliation. They enforced an embargo on all food transport into the western regions, and the hunger began to set in. Butter ran out in October, and animal fats days later. The Germans had removed large quantities of cattle and other food stocks too. The embargo was partially relaxed in November, when Berlin allowed restricted waterborne food transport, all so important in the canal-strewn Netherlands. But winter was coming, and by the time the waterways were given the green light, they were already frozen solid, and the Hongerwinter began.

For the next few months, famine struck hard. The best estimates are that more than 18,000 people died as a direct result of malnutrition. According to contemporary sources, tulip bulbs became stock dietary supplements, as did sugar beet. Fuel ran out, and the

Dutch ransacked wood from buildings and tram tracks to burn in stoves. This was already a time of rationing, but with the food supply shrinking to almost nothing, the portions became smaller and smaller. Allocation of vegetable oil was restricted to 1.3 litres per person for the period September to March, that is, one small cup per month. Potatoes were rationed too, but soon there were none, and potato vouchers were exchanged in communal kitchens for watery soups. The bread ration had fallen over the previous few war years from 2,200 grams per month to 800 grams in November, and this would halve again in April 1945.

On Sunday, April 29, the RAF and the Royal Canadian Airforce enacted Operation Manna, and the USAF Operation Chowhound, two mercy missions that the Nazis agreed not to shoot down, as they dropped food into the famine. Six days later, the Reich capitulated, and the food began to flow back into the Netherlands.

Because this horror happened during the modern age, with record keeping and an understanding of public health, and also during the period of rationing, the Nazis had effectively executed an experiment that we could or would never do. They had asked the question: "What happens to people when you profoundly starve them?"

Many of the results, though tragic, are not very surprising. People who lived through the famine suffered a collection of health problems, from the physiological to the psychiatric. Audrey Hepburn was fifteen at the time and endured the Hongerwinter like so many others, making flour for bread and biscuits out of dried tulip bulbs. As an adult she described surviving the starvation with anemia, respiratory problems, and edema, and attributed health problems throughout her life to the malnutrition she endured in 1944.

These conditions are all miserably predictable. Their DNA was already set in place. But the unborn would face a different set of problems. For children in utero, maternal malnutrition has an effect on the expression of their genes, which would be altered and disrupted. The result for the children of the Hongerwinter was a set of health problems that would persist throughout their lives. Compared to the general population, and compared to siblings who were conceived either before or after the famine, they were smaller and underweight. They were more likely to be obese, diabetic, and schizophrenic, and to suffer from cardiovascular diseases and breast cancer.

Malnutrition can affect the body in many ways, but one of the ways the external world interacts with our inborn biology is via the domain of epigenetics, a word that literally means "in addition to genetics." DNA does not change in a person's lifetime (unless affected by random mutations from emerging cancers, or caused by attacks from radiation or other mutagens). But with all DNA being present in all cells, mechanisms are needed to make sure only the right genes are on at the right time. Epigenetics is one such system of gene regulation—a way of modifying DNA without altering the gene sequence itself.

Think of DNA as an orchestral score, the notes on the page unchanging. If you buy the sheet music to Beethoven's Seventh Symphony, it is the same notes whether it was published in 1816, 1916, or 2016. How it sounds when played by a full orchestra is another matter though. A conductor and each player in an orchestra will interpret the notes and place annotations on the manuscript to dictate how the music sounds, with crescendo and diminuendo and adagio. Each performance is unique, even when the original scores are identical. In DNA, the notes never change, but they get annotated in all sorts of ways. The most common is via small molecular tags called methyl groups; it's simply a carbon atom attached to three hydrogens, and these stick in humans onto the nucleobase C (cytosine) in DNA. This has the effect of silencing that bit of DNA. If it's in a gene, that gene cannot and will not be transcribed into RNA, and will never become protein. It is still there in the genome, but has been scored out. If it's attached to a promoter of a gene, then that gene will not be activated, an instruction akin to "don't bother with the following sentence": It doesn't help the story.

The DNA sequence itself does not change—nature—but the modification of DNA does and is reversible according to the actions of the organism—nurture. We've known about this for decades, the most obvious example being carried by just over half the people on Earth. Women have two X chromosomes, but only really need one. So, in every cell, one X—selected randomly—is permanently labeled with epigenetic methyl tags, which effectively redacts the whole chromosome. Many individual genes are modulated like this too, and many corresponding traits are dependent on this system. Rat mothers lick their pups, and those that are licked less have measurably higher stress levels, which correlates with less epigenetic tagging on genes associated with stress. Some breeds of mice can be given foods that alter epigenetic expression of a gene that is involved in their coat color, which also has the knock-in effect of making them heavier and more prone to cancer. That sandwich you just ate, that triggered some epigenetic changes in your body as a way of modulating what genes you need to be active over the next few minutes and hours as you digest it. Epigenetics is a very important part of normal biology and an area that is blossoming, as new techniques are being developed to measure how the tagging takes place and in response to environmental triggers.

But, as sometimes happens in science, it's also a field that has been exposed and leapt upon in a press frenzy to attempt to explain all sorts of as yet unsolved mysteries of biology. The legion purveyors of ackamarackus* love a real but tricky scientific concept that they can bolt their quackery onto. It happens with words like *quantum*,

---

* My editor insisted that there was not room for two flapdoodles in one book,

which offers up some magical scienceyness, none more so than in *quantum healing*—an unfathomable extension of reiki, which, let's face it, is a load of old cobblers already. The annexing of this word from fundamental physics also ranges from washing powder branding to the theory of mind. Lots of real scientific terms get borrowed for a spot of buzzword scienceyness. We see the same effect with the words *neuro-* or *nano-* added to almost anything. Neuromarketing, neuroentrepreneurism, neuropolitics are all new fields in which flaky science is used to put a patina of credibility onto a product. Due to predictable quackery, possibly embiggened by some overstatement or questionable scrutiny within science, epigenetics is threatening to become the new quantum. It has been speciously applied to all manner of healing, as if it were magic, when in fact it's just another bit of biology.

That hideous Nazi experiment in the Netherlands generated an incomparable dataset of the people they imposed starvation on. The people of the Hongerwinter have been studied throughout their lives and the health problems they were issued with by the cruelty of the Reich have provided vital information about the effects of malnutrition. What was unexpected, though, was that some of the consequences continued into their children. People who were conceived during the Hongerwinter are now in their seventies, and many of them had children who are also now adults. It was not startling that so many of the survivors had so many health problems, but the fact that their children did too was a surprise. In mammals, epigenetic modifications tend to get reset each generation, but some, very limited, rare epigenetic tags appear to be passed down from parent to child, at least for a couple of

---

so here is an equivalent word for tosh, baboonery, claptrap, flimflam, rannygazoo, baloney, or just plain old nonsense.

generations. We've seen a handful of these in mice, even fewer in humans. In rats, learned behaviors such as fear or stress can be passed on to their children, and subsequently to grandchildren, also by proposed epigenetic mechanisms. This means, as far as scientists can tell, that DNA methylation of certain genes occurs in the mothers, is not reset as we would expect, and persists after conception of their pups.

The children of the Hongerwinter babies were not smaller, nor did they have an increased chance of cardiovascular disease. However they were fatter, or rather, they had increased neonatal adiposity. Later in life, this correlates with increased risk of diabetes and other health problems. This result was a first in humans, that showed that what happens to a mother can affect not only her children, but her grandchildren too.

There are but the smallest handful of similar results in people. One frequently cited study concerns the population of the rural Swedish district of Överkalix, which over the last century or so has been subject to highly variable harvests. Life expectancy was significantly raised in men whose grandfathers had endured a failed crop season just before puberty: They had acquired something due to starvation, and passed it on. A similar result from Bristol scientists using a huge dataset called the Avon Longitudinal Study—the gold standard of transgenerational research—showed that men who smoked before puberty sired fatter sons than those who smoked after. Again, something was apparently being acquired and passed on.

These results are complex, perplexing, but possibly slight, and demand more work and greater examination. Science is unfortunately prone to fashion, and many scientists are intrigued but anxious that the scrutiny being applied to the studies of transgenerational epigenetics are not robust enough to justify the fanfare. That applies to the experimental studies in rodents as well as the fascinating but puzzling observations seen in the tiny smattering of transgenerational genetic effects in the Dutch and Swedish starvations.

One of the reasons that people get excited by these discoveries is that, after Darwin, we rejected the notion that evolved traits could be acquired during life. Jean-Baptiste Lamarck is the figurehead for this idea, and it bears his name to this day. When Lamarckian inheritance is taught it's often mocked for being wrong: His idea was that use of a trait would promote its transmission to subsequent generations, and disuse would reduce its transmission. Actually, Lamarck was a terrific scientist, and contributed many great ideas to the developing science of biology in his eighteenth- and nineteenth-century lifetime. But on evolution via acquisition and practice he was wrong, as good scientists often are. Darwinian evolution, via that blind mechanism of natural selection, superseded Lamarckian thinking, but only in the analysis of his ideas after he had died. In fact, Darwin himself was effectively in support of a Lamarckian view for some characteristics. He postulated a hypothesis in 1868 that non-sex cells in plants could acquire environmental cues, and as a result throw off "gemmules" that would aggregate into the germ cells, and thus the plant would produce offspring that had responded to their parent's life experiences. August Weisman tested a similar idea in an experiment in which he cut off the tails of sixty-eight mice over five generations. Of 901 pups born, none of them was born without a tail. He wasn't really testing use of a trait, but nevertheless was asking whether evolution might follow an acquired trait. Of course, as geneticist Steve Jones enjoys pointing out, the Jews have been performing a version of this experiment for a few thousand years, and so far a boy without a foreskin is yet to be born. Lamarckian inheritance was an idea, and it was wrong. The genes are set in place in each parent before conception of their offspring, and changes to DNA that result in evolution of species will have already occurred by conception.

The transgenerational studies in mice, rats, and in those very few human experiments at first glance appear to contradict this hard core of evolutionary theory. But do they? The answer is probably not. If the changes are permanent, then we do have a puzzle. But given that in mice the changes have at best only lasted a few generations before fading, the effects appear to be intriguing but not revolutionary. We don't yet know about the permanence of the effects in humans, we're just annoyingly slow to grow like that. Creationists (and others unencumbered by facts) cite epigenetics to assert that Darwin was wrong, and that these transgenerational epigenetic studies show Lamarckian evolution. They don't, as the changes are not perpetual and do not change the DNA sequence itself, on which natural selection acts. Even the few inherited epigenetic changes we observe are not very predictable, let alone predictably positive. The Överkalix grandsons lived longer if their grandfathers lived through famine. But the granddaughters of women who had survived fallow seasons had lower life expectancy. Conclusion? Inconclusive. More work needed. Even if, one day, we did show that epigenetic tags were permanently heritable and therefore could be selected, it would still only be a drop in the evolutionary ocean. Show me one robust example, and I'll show you literally a billion that are straight-up Darwinian.

New Age gurus and gullible journalists cite epigenetics as a way of changing your life, under the false supposition that genes are destiny, and epigenetic changes brought on by lifestyle choices such as meditation "allows us almost unlimited influence on our fate," to quote the supreme New Age guru, Deepak Chopra.

I suppose that really depends on what you mean by *fate*. If you are fated to digest your lunch, then yes, epigenetics will play a key part. If you are destined to go to sleep tonight, your epigenetic tagging will change accordingly. There will be other, perhaps less trivial epigenetic modifications during your life. Methyl tagging can affect genes that are involved in cancers, and indeed drugs are being developed that reverse epigenetic tagging in order to reactivate erroneously muted genes. All these things are important parts of biology that deserve proper scientific scrutiny in the years to come. When it comes to the transgenerational effects seen in a few, slight studies, let's wait and see if these effects are real, and how strong they are.

In the meantime, let's recognize epigenetics for what it is: a fascinating, essential part of biology, still in its infancy, worthy of serious, scrutiny-rich research. It is not magic or new, not heretical, and it won't upend Darwin or gift you supernatural powers over your life and fate. Epigenetics is a necessary pursuit in our never-ending quest to unpick the inscrutableness of being. The invention of techniques that enable the significance of more and more epigenetic effects to be understood is an exciting emerging field. That work is needed, unhyped and solid. While we await the solving of those mysteries, we would all benefit from remembering that mystical thinking is never welcome in science.

8

A short introduction to the future of humankind

"Greetings, my friend! We are all interested in the future, for that is where you and I are going to spend the rest of our lives! And remember, my friend, future events such as these will affect you in the future."

The Amazing Criswell, *Plan 9 from Outer Space* (1959)

A television producer once took me to lunch to ask me a very important question: "When will humans evolve the ability to fly?" A new superhero series was coming to television, with mutations in the characters' DNA that were gifting them uncanny comic-book powers of flight, telekinesis, time travel, mind control, and so on, rather like Marvel's mutant X-Men, but definitely not Marvel's mutant X-Men. The producers were interested in a science program that might sit alongside the drama, about the real science of incredible human abilities, and the real-world possibilities of the evolution of these types of superpowers.

My answer was quicksilver fast: We already have. I love comics, and read them to this day, and have spent (arguably, wasted) much of my free time over the last thirty years considering the possible realities of superpowers. But the answer doesn't rely on strange tales or amazing fantasy. So, I made a great grandstanding speech about how we have evolved massive creative brains, capable of planning and predicting the future, of invention and creativity, and this had helped us extract ourselves from many of the historical shackles of natural selection. We have externalized the stomach with the invention of cooking, so we don't have to digest a whole range of chewy molecules, because they are already partly broken down by our unique control of elemental fire. We have bypassed many aspects of a life of nomadic sustenance, as well as hunting and gathering, by settling and domesticating all manner of beasts of the field and plants of the ground. This also has changed our culture, technology, and even our genes (as discussed in Chapter 2). We have radically eliminated diseases that scythed down ancient populations with casual indifference—plagues, malarias, cancers, pestilence. Smallpox once killed hundreds of thousands every year. Since the 1980s, as a result of vaccination, there have been no cases of smallpox. Polio looks set to follow soon as a disease only of interest to historians. These sorts of evolutionary pressures have been radically altered as a result of invention and science and the technology that has come about through our own evolutionary trajectory.

"How long before we fly? We do it all the time." I waxed lyrical. We invented airplanes and helicopters, and rockets to explore space, and even hoverboards and jetpacks are not far off. We've walked on the Moon, and soon, a son or daughter of this planet will walk on another, just as Kal-El, the Son of Krypton, came to Earth. "We are superhuman already."

He looked pleased, and impressed. And then said, "So you think we'll evolve how to fly within the next couple of centuries?"

I finished my pizza, thanked him, and left. Limbs do indeed come and go in evolution. The basic body plan of all animals is broadly similar, which betrays a common evolutionary origin. The genes that indicate that "a leg should go here" are broadly the same in all species with legs, meaning they have deep roots. These, called Hox genes, specify what parts of a body will go where, and it is in their mutation, duplication, and multiplication that the variation in bodies in almost all creatures resides. Insects generally have six legs, and spiders eight, but millipedes and such creatures that crawleth have loads. But they have simply arisen from the duplication of the genes that issue the standard six in the majority of small arthropods.

Bat wings and bird wings are functionally the same, but have come via different evolutionary routes. We call this "convergence." A hundred million years ago, the forearms of some dinosaurs had shifted their form to allow gliding, and their bones became lighter and hollow. Dinosaurs had feathers already long before they had flight, including the mighty *Tyrannosaurus rex* and even bigger, scarier, yet perversely fluffier predators, but over millions of years they would become part of the process of uplift and propulsion through the air. Mammals had split from dinosaurs epochs earlier, and the weeny critters that would form the basis of 1,240 bat species weren't airborne for tens of millions of years yet to come, and have never had feathers. Structurally they are similar but different; they're both adaptations of the forelimbs, meaning that their common root is not in flight but in the origin of four limbs. As forelimbs they are homologous: Equivalent bones are present, stretched and contorted into their distinct morphologies. In providing the uplift, flap, and glide of flight, the wings of birds and bats are—in the technical language of evolutionary biology—not homologous but analogous, as indeed they are with insect and airplane wings. Unlike the functional similarity with insects, though, both birds and bat wings have similar bones, as indeed both do with dolphin fins and horse's hooves, again showing the four-legged root of these appendages. But the winged portions of their forelimbs are different, showing independent evolutionary origins. So to see evolution of our own forelimbs into wings would see the relinquishment of our hands and arms or the growth of an entirely new set of limbs. There isn't a great deal of evolutionary pressure for either. More importantly, there isn't a gene "for" wings, and we would have to undergo a grotesque, energetically demanding transmogrification in utero for arms to become wings, and that couldn't and wouldn't happen.

Bats and birds have acquired their flight over thousands of generations of slight incremental change conferred by subtle shifts in many genes being selected as they provide an advantage to the owner. Because limbs are so ancient in the grand scheme of evolution, I suppose, if we gave in to some fantastical thinking, that an unborn might harbor a genetic anomaly that would stimulate growth of an extra pair of limb buds. That child would have to successfully reproduce and pass down the anomaly, and over generational time the genes that prompted it would have to spread through the population, each time being slightly selected to be bigger and more winglike. It would, I guess, take dozens, or hundreds of generations. Wings are merely vehicles for providing the best chance of propelling some genes into the future generations. In birds, as soon as wings stop providing that propulsion they are gone, as we see in the flightless emus, kakapos, and kiwis. And insects have acquired and shed wings seemingly willy-nilly in their evolutions as they confer advantage or not for the genes they carry. They are to evade predators, or to gain access to food, or to show off to the females—all pretty standard reasons for having any evolved trait. The timescales involved are also unimaginable to us. So while, hypothetically, morphologically just conceivably possible, the chances of this coming to pass are about as likely as a boy bitten by a radioactive spider gaining the ability to produce webs from his wrists.*

---

* In the original incarnation of *Spider-Man!* (*Amazing Fantasy* #15, 1962), Peter Parker invented wrist-based web spinners, as he was a teenage science

But aside from the physical unlikeliness of the sudden acquisition of wings, and the huge metabolic demands of growing limbs powerful enough to let loose the shackles of gravity unaided, the real reason we will never evolve flight is the one I gave to the producer. We fly all the time. We have no need to fly unaided. There is no ecological niche that could be filled by aerial people. If by some incomprehensibly reality-defying mutation a child was born with the nascent power of flight, their advantage over us earthbound dawdlers would be negligible, and their freakishness would probably render them an unlikely sexual partner.

This isn't how evolution works. It's an ultra-simplified dilution of evolution drawn from comic books, science fiction, and creationists. We see in all the wondrous adaptations of us and all species a fit for purpose. It's so easy to apply that adaptationist observation that all things are there to suit the things they work best for. We don't have prominent noses so that we can balance spectacles on them, as Voltaire's Dr. Pangloss suggested, but as a result of history, time, chance, and sex. Noses and indeed smelling predate humans by hundreds of millions of years, so we carry that ancestral burden; over eons our noses have changed shape via natural variation according to what use they provide, the overall changing shape of our faces, and what we find attractive. The morphology of our noses is indeed underwritten by genes, and it is the changing frequency in a population of different alleles that is the measure of evolution. Nose shape is perhaps a bad example, because I believe people with big and small noses reproduce with equal success (at least, as far as I know; studies have not been done on nose size and reproductive success). The subtler questions of alleles that reduce or enhance reproductive success, or the health that

---

genius. Comic book superheroes reinvent themselves every few years, and he acquired biological silk production only in the 2002 film—a teenage boy emitting white sticky fluid as some sort of allegory for something or other.

underlies that, are really what's at stake in the future evolution of our species. "Are we still evolving?" is a question that geneticists get asked a lot. Here is the answer: yes.

Our genomes are where evolution takes place. Our DNA changes over time, every generation. Most of these changes are subtle, many trivial. Some are teasingly interesting. We humans are trichromatic—we see in three colors. In the back of our eyes we have photoreceptors, highly specialized cells whose purpose is to literally capture the photons of light that flood through our pupils. There are two classes commonly known as rods and cones: The rods are attuned to pick up movement and low lighting conditions, and they sit in the periphery of the retina, which is why we see indistinct but moving things out of the corners of our eyes. The cones are central, which is why your sharpest color vision is right in front of your eye. If you wave something in a hand far outstretched to your side and look straight ahead, you can see it move, but not what color it is. Then there are three types of cone, each further attuned to a specific wavelength of light, which determines what colors we see. Broadly, they are short, medium, and long wave, but roughly correspond to blue, green, and red, though they overlap in their range, and are subtly variable between people. The difference between each of these cones is down to a single protein called an opsin. The photon passes through your clear cornea and the nucleus-free cells of the lens, through the jelly aqueous, then vitreous humors, through three layers of brain cells, nerves, and blood vessels, and into the very back of the eye where the opsins sit bound into the pointy tips of the cones.* There, the photons are captured

* It's an odd way to design an eye, to have layer upon layer of obfuscation before getting to the actual bit that does the work of capturing light, the photoreceptors. It's like holding the fuzzy end of a microphone away from the person you're trying to record, and pointing the wire at their mouth. But then evolution, ironically, is blind, and what works, sticks, at least for as long as it

by the opsin molecules, which physically jiggle their shape in response, and that molecular shrug triggers an electrical impulse, which shoots out of the other end of the photoreceptor and through the several layers of nerve cells, which collectively bundle their nerve fibers into the optic nerve, into the visual cortex of the brain, and this is how you see.

Many mammals have only two cone opsins, and so see color with less acuity than us. Most apes have three, as do the Old World monkeys that are indigenous to Africa and Asia. Cats have many more rods and so see in the dark much better than us, but not color. Certain species in the family of the mantis shrimp have at least sixteen opsins, fine-tuned to see red, blue, and green, as well as polarized light, ultraviolet, and a host of light unseen by us that we can only dream about.

The mutations that gave rise to three colors in us (and the many in the shrimp), were not initially the single letter changes that make up most genomic change, but large duplications of whole sections of DNA, and subsequent typos. Colors are determined by the wavelength of the light we see, and the gene for the Shortwave opsin is on chromosome 7, whereas the Medium and Long are on the X. This is why men are more prone to color blindness than women: A faulty opsin on one X can be compensated for by a

---

does work. Mexican cave-fish lost their eyes because they're not useful in the environment they live in, which is as black as the night. Reversing our backward retina would be a colossal genetic task, and in a similar way to the unlikelihood of us growing wings, probably will never happen, though octopus and squid do indeed have their retina the sensible way around. Of course, the eye is frequently the weapon of choice of creationists in the eye-rollingly tedious ongoing battle between fact and those unencumbered by fact. "The eye," they bleat, "is too complex and perfect to have evolved by chance!" A squid might agree, but try telling that to someone with cataracts, myopia, strabismus, microphthalmia, aniridia, presbyopia, detached retina, glaucoma, color blindness, or any of the dozens of design flaws that are either a result of evolution's lack of foresight, or a really rubbish designer.

woman's second; men have no such insurance. The duplication of one opsin on the X to two at some point in our primate evolution allowed one of them to mutate freely without a loss of function, and thus we were free to acquire a new color sensitivity. That all happened tens of millions of years ago, long before humans, but something similar might be happening now in us—in fact, some of half of us. Some women might be tetrachromatic. They, through another random chance duplication, have acquired a fourth opsin on one of their X chromosomes. Around one in eight women are estimated to have this extra gene variant, but whether that bestows tetrachromacy is not yet known. The ones who do have this power see colors where we see monotones. It's a new area of research, and the condition appears to be rare, and poorly accounted for. A few women have been studied, and they seem to see clear differences in colors that are merely shades to normal trichromats. When examining red-green color blindness, the Ishihara test presents a circle containing circles in different hues. Hidden in plain sight (to those with typical vision) is a number, but due to the design of the shades that pick out the number, it is invisible to color-blind people. The tetrachromat tests also rest on the ability to discriminate distinct hues of green where we only see olive.

The theories behind why we evolved three-color vision are wide and varied. Many of them suppose that the ability to discriminate the redness of berries in a busy green forest canopy would be of great advantage to our foraging simian ancestors swinging in the trees.

The advantage of the ability to discriminate four colors is a mystery. While many animals have more than our three, tetrachromacy in humans is likely to be recent and random—chance plus time—but not a mutation that has been negatively selected as it is unlikely to cause any phenotypic problem. It simply is—another example of our infinite variation. It's not likely to spread far and wide, but who knows? Ask me again in 5,000 years.

This is just one example of a mutation that has emerged that has a distinctive effect. Most do nothing of such interesting significance. DNA replication is imperfect, and we have plenty of spell-checkers in our cells. But mutations arise, mostly single changes, occasionally chunkier sections of DNA. Without them, no evolution would occur, for there would be no variation on which selection could act. From an evolutionary point of view, perfection is boring and impractical, and infidelity is essential, at least when it comes to the code. The process of DNA replication has to be imperfect. You have acquired through nothing other than chance at least 100 mutations that are unique to you. If you have children, you may well pass them on to your children, and they will acquire plenty of their own. As long as humans keep having sex and that sex results in more humans, then we are evolving. We can avoid these evolutionary changes no more easily than we can change the weather.

Of course, we have profoundly changed the weather. We farmed the land for 10,000 years and hunted animals to extinction. Our presence on Earth has altered its terrain along with the fauna and flora. Climates changed over epochs and we adapted. Ice ages came and went, and, nowadays, they may not return as a result of our own action. Through modern farming practices and through less than three centuries of continuous industrial revolution we have made the earth considerably warmer. During that time our lives have become unrecognizably healthier. Life expectancy is hyper-variable across the world, indeed across Britain, and even decreases measurably by Underground station as you travel from affluent central London out east into the relative poverty of the suburbs. But across the board, life expectancy has increased, on average and in real terms. We have fewer children than at any time in history, and more of them survive. In most cultures, we tend to marry whomever we wish. All these things might point toward us being unshackled from the binds of natural selection. We are free to choose with whom we partner (including, in more and more cultures, members of the same sex), when and if we have children.

In the recent past, though, our farming changed us: in the food we eat, in the genes we carry to process that food, in the milk that we drink, in the northern climes that we moved to. We've cleared forests, which brought still water and mosquitoes to breed in it, and we've evolved in response to that too, with the protection of sickle trait for carriers, and sickle cell anemia for homozygotes.

The new world of genomics has provided a colossal data set from which we can effectively compare all humans to see the rate of evolutionary change in our DNA, not just looking at individual genes that have popped up and given us new powers or new traits, but across our entire DNA. We use reference genomes, and panels and databases of variants, common and rare. In 2013, Josh Akey and his colleagues from the University of Washington in Seattle compared the genomes of 6,500 people, just looking at individual variations in SNPs. They scanned 15,000 genes, and found 1.15 million SNPs. They applied six different tests to gauge when each of these alleles arose in history, including factoring in that our numbers as a species have increased more than a thousandfold since we started farming. Of these differences, three quarters had arisen in the last 5,000 years. Are we still evolving? The answer again is an undeniable yes: We're a species not of mutants but of mutations.

Approximately 164,688 of these single changes in our DNA are probably not good news. They're changes to the DNA that alter proteins subtly, but probably making the proteins less efficient or worse, dysfunctional. According to Akey's data, 86 percent of these also arose in the last 5,000 years. We are indeed evolving, and we are acquiring new genetic problems.

This is perhaps not surprising given our global creep from Africa. When that happened, some 100,000 years ago, only a few thousand individuals stepped forward to be the progenitors of the rest of the world. They went forth and multiplied. Five thousand years ago there were 5 million or so of us. By 2025 we estimate a global population of 9 billion. They spread in all directions, into Europe, into Asia, across the Bering land bridge, and down all the way through the Americas, east to China and south to India and Oceania.

Those migrants left their genetic traces wherever they went. We've seen how farming cultures changed our genomes, we've seen how the presence of diseases did the same. Culture has shaped our changing genes. Today, if you go for elective surgery in Hyderabad in India, one of the first questions you will be asked is "Are you a Vaishya?"

Vaishya are one of the four principal castes in India, known as a "merchant caste," in comparison to the Brahmin religious caste or the Untouchables. That question of social identification sits at the top of the consent form in Hyderabad hospitals. Some assume that the question demonstrates latent social prejudice, which is endemic in India. In fact, it is a smart question rooted in the evolution of the Indian genome. In the 1980s, surgeons began to notice that some patients remained unconscious under anesthetic for several hours longer than they should. Typical general anesthetics are a cocktail of drugs that do specific things to beckon the sandman. By a process of elimination, the doctors identified the cause: A short-lived muscle relaxant called succinylcholine used to suppress the musculature of the lungs so that doctors can feed in a breathing tube without them closing shut in an attempt to prevent invasion. In certain cases the effect lasted for hours, rather than the typical minutes. There were no serious health repercussions, just a prolonged period of artificial ventilation and a weirdly long sleep after a minor operation. On closer inspection, they discovered that this only occurred in Vaishyas. By looking into their genomes,* we learned of a single change—a random switching of a single letter of the gene encoding the enzyme butyrylcholinesterase (BCHE), which normally helps degrade molecules in the blood similar to the anesthetic.

Upon further inspection it turns out that Persian Jews and Inuits are also prone to the same pseudocholinesterase deficiency via a different mutation. According to Indian geneticists in Hyderabad, the mutation had occurred in an unknown person around a millennium ago, and it was preserved within that group of people for evermore by endogamy. The Vaishyas are more than 20 million in number, so it seems unlikely that large-scale inbreeding could have maintained the variant within such an expansive group—even caste is not so watertight as to prevent such genetic leakage. Others have speculated that this particular allele might have conferred some incremental advantage in an unknown condition, like heterozygosity for sickle cell is protective against malaria. Vaishyas and Inuits have particularly fatty diets, ghee and blubber being the sources, respectively; as a result both populations are frequently obese, and maybe this variant has some role in the metabolism of

* The survey of the Indian genome began in 2009, sampling people from up and down the country, from isolated villages to the big cities. It revealed that despite the contemporary population's incredible diversity, it is derived from just two distinct ancient populations. One of these, from the north, were distant cousins of Europeans and Middle Easterners, who first went north out of Africa. Those from the south were as different from the northerners as they were from the Chinese. These distinctions are not visible now: the two decks have been irreversibly shuffled. Almost all Indians sampled showed a blend of those two ancestral groups, but in differing proportions. The genomes revealed a range of genetic diversity in India up to four times greater than that found in Europeans: Indians of different groups are less similar to each other than a Scot is to a German.

these heavy saturated fats. Traces of selection in and around the butyrylcholinesterase gene haven't shown any signs of selection so far, nor is there evidence that it has surfed along with neighboring genes that have provided an advantage to the owner. Nevertheless, this version may have simply had no effect for centuries, just a typical SNP in an inconsequential allele. With the advent of modern medicine, this natural variation suddenly became of interest. Now, their anesthetics are tailored to their genomes, evolution molded by the culture of a people. That allele will simply persist now. It has evolved, but may well have been neutral for all but a few years of its existence. Though now it has an effect significant for medicine, it's not lethal nor does it reduce reproductive fitness, and so is not subject to the pressures of selection.

Caste is a strange evolutionary experiment. It's a social hierarchy of immense complexity that primarily prevents marriage between individuals from different social groups. Despite modern attempts to ease caste boundaries, and though in urban centers the rules are relaxing, many if not most marriages in India are still arranged. The "would like to meet" and "good sense of humor" biographies in personals ads in newspapers are still filtered by caste. This means that genes and genomes have been—and continue to be—kept largely within discrete groups of people for generations. In countries like the UK, where an historically blurred social structure evolved, intermarriage between social groups is much more common, and interbreeding between the traditional upper and lower classes has occurred with the vigor of Lady Chatterley and her low-born lover. The houses of the elite rise and fall. In India, the social restrictions on caste are more strictly endogamous. For a time, there was a suspicion that the caste system was locked down in India during British colonial rule—that the British enforced and encouraged a preexisting, more informal caste system as a means of social control. With genomics we can make an estimate of the point where a social convention impacted upon the Indian genome. Studies on the genome in 2013 revealed that alleles of caste began to become stratified via endogamy at least 1,900 years ago, a date different from and far more precise than that suggested by the blurred lines of history. DNA says that caste predates colonial India by centuries.

The evolution of the genome is unequivocal. We shape it with culture and with technology. We have shaped it via our voluntary isolation from our African roots and subsequent explosive multiplication. It continues to change every generation. The question is not, "Are we still evolving?" It is, "Are we still under the spell of natural selection?"

That is a harder question to answer. Much of its inherent difficulty lies in the word *natural.* Nothing of our lives could be considered natural in any meaningful sense. We have altered every conceivable aspect of the world in which we have lived for tens of thousands of years, and in controlling (or at least attempting to manipulate) that environment we've profoundly changed the grip with which evolution would have tenaciously selected the differences between us. Food and sex are the two key resources that propel our genes into the future, but we invented almost all our own foods via that ancient form of genetic engineering better known as farming, and nowadays we eat what we want. We sleep with who we can, and almost never for the production of small humans. Sanitation, housing, medicine, and wealth have inoculated and insulated us from the pressures that historically would've cut through populations. Most women now live to an age when they can reproduce, and most do. On the whole, we have as many children as we wish to have, not as many as possible to maximize the chances of our genes surviving the assault of existence and time. In the nineteenth century, women in the UK averaged 5.5 children, but by the end of the First World War it was down to 2.4.

These are two key factors in assessing whether our ongoing evolution is undergoing a form of selection: the death of babies and the number of babies women have. Scandinavia is the territory from which we have the most data: In Sweden in the late eighteenth century the child mortality rate hovered around one in three. Today it's three in a thousand. This difference is part of the changing face of humankind's relationship with evolution. This release that we have invented from the harshness of nature means that evolution with selection is likely to have radically slowed, though not stopped. The number of children who die is a driver of evolutionary change, as their genes will not continue, and the frequency of alleles that survive are much more likely to propagate down the generations and into populations. By reducing that number via vastly improved medicine and public health, and contraception, the purchase of selection on humans is reduced.

The potential balance comes from choice. The period during which women can bear children has increased, as has lifespan, but the time during which they do has changed and continues to do so. As long as there is variability in the number of children that women bear, and as long as there is variability in the number of babies that survive themselves, then there is the potential for selection to act. These are questions only recently asked, and data that has not been collected for long. Infant mortality is highly variable through history, and though it is generally falling, is still highly variable all over the world. The death rate of children in rich countries has decreased radically, and the data over the last century is precise. According to the United Nations, the infant mortality rate today for many wealthy countries (Singapore, Japan, all of Scandinavia, much of Europe) hovers around two or three deaths per 1,000 births, defined by death of children before their first birthday. The UK is 4.19 and the USA 5.97. The bottom ten are all in Africa, between 70 and 90 deaths per 1,000. In 1950, Scandinavia was still top, but the number of deaths was around 20 per 1,000 for Sweden, Norway, and Iceland. Only six of the bottom ten were in Africa. But their numbers were around one in four (250 per 1,000).

All this inequity makes painting a global picture of our species' evolution near impossible. We are too widespread and inequality reigns supreme. Collecting the numbers on infant mortality plays an essential part, but when what we are really interested in are the changing frequencies in bits of DNA, there's just not enough data. Only in the last few decades have we had the foresight (and funding) to do scientifically what family trees have done for thousands of years—trace people through time, generation after generation.

A long-standing study of the population of the pretty Massachusetts town of Framingham has been running since 1948, initially just collecting data relating to heart health. But tracking more than 14,000 people over time is a seam of such potential riches that over a thousand papers have been published from doctors and scientists observing the people of a town for fifty years. What they see is a fluctuation in the number of children that women bear, and this relates to culture. In the 1950s and '60s, when smoking, exercise, and fatty foods were of less concern to us, the Framingham women had, on average, fewer children than in the 1990s. For evolution to proceed, the traits need to be heritable. While smoking and high cholesterol levels have heritable components, this is not the potential engine for this possible evolution. It is simply that by living healthier lives, the women of Framingham were having more children, and thus passing on more genes into the future. They started, on average, having children earlier and finished, on average, later, meaning that they could squeeze out another baby, on average. They also found that, on average, slightly plumper and shorter women had more children. If the trend were to continue, unfettered by any change in the local cultural environment, then by 2400, the women of Framingham would be around an inch shorter and two pounds heavier, on average. That sort of slow demographic shift is typical for evolution more broadly, but not exactly revolutionary. It could also be wiped out by a sudden shift in environmental factors—a change in school meal policy, or something as seemingly trivial as a factory closing and a migration out of town. Traits never evolve in isolation from the environment or each other.

The Finns have scrutinized their own evolution most recently in the most depth. In 2015, scientists compiled data from 15 generations and 300 years of genealogical records—around 10,000 people. Three centuries covers a shift from an agricultural and fishing culture to an industrialized nation, and all the cultural mores that come with those revolutions. The number of Finns in the eighteenth century was merely 450,000, recovering from being a ransacked battleground for the victorious Russians to the east and the Swedes to the west. The Russians came and went, and came again, and the Swedish did the same. Sporadic peace and prosperity resulted in the population doubling by the nineteenth century, and by 2010 it was more than 5 million. Their birth rate had dropped from 5 in the mid-nineteenth century, to 1.6 per family today, but the survival rate more than compensated. Two thirds of Finnish babies survived in the 1860s, but by the time of the Second World War, that number was greater than 94 percent.

Elisabeth Bolund and her colleagues pored over church records and reconstructed the most comprehensive family tree in Finnish history. They disentangled meaningful data on lifespan, number of children, and age of the mother at the time of her first and last child, and by looking at how these numbers panned out in families over time, they could extract what proportion was due to genes, and what to the circumstances in which they were operating. Bolund found that between 4 and 18 percent of the variation in the birth data could be attributed to DNA. But they also showed that its influence has increased over time. It may be that with increased access to healthcare and sanitation, the effect of the environmental inequity is reduced, and so the impact of genes rises, at least in the trends of births and pregnancy. This, in principle, gives more purchase to Darwinian selection.

Whether this effect is felt elsewhere or worldwide is unknown. But the question of current human evolution relies heavily on two factors. The first is the question, "Is variation heritable?" To this the answer is yes. We see it in specific genes, and we see it accounted for in complex behaviors for which the specifics of DNA are not necessarily known. The second question is, "Does the survival of babies vary?" The answer to that is also very measurably yes.

The timescales in these studies are minuscule, and the changes small. The chasms between infant mortality rates around the globe means that these evolutions in affluent populations like Finland and Framingham cannot be extrapolated nor generalized, and while we see slightly similar results in small studies from the Gambia and elsewhere, the data is far from comprehensive. A few generations is a timescale in the evolutionary shallows, and we can only dip our toes in the observable data. But we are evolving, our genomes are changing and though the pressures of selection have been radically altered, humanity's grip on modernity has not removed them altogether. We are not creatures fixed in time, we remain a species begotten not created. As long as there is difference, we will always be a species in transition.

The history of everyone who has ever lived is buried in DNA in us or in the ground, but it's tough making predictions, especially about the future. If this all sounds a bit inconclusive as this book draws to its end, that, happily, is the nature of science.

> Ignorance more frequently begets confidence than does knowledge . . .

says Darwin in *The Descent of Man*, his study of the evolution of humans from our hairier ancestors. It's no secret that arguing with creationists is a waste of time, for they see things differently from most. They know that what they think is true, and in science we must assume we are wrong. We know that we doubt everything we know is correct. When it becomes harder and harder to find things wrong with your ideas and experiments, then it's probably a good sign you're on the right track. That is science's eternal trump card. Christians sometimes say that they perpetually doubt their own faith, but it's a different type of doubt from the scientist's. Christians appear to doubt under the assumption that their assumptions will be affirmed. Scientists doubt with the explicit plan that their results will be overturned. Most Christians I know also think Darwinian evolution is the best description of how things are the way they are. Nevertheless, creationism does exist as a fringe, yet vocal, branch of Christianity, frothing with risible fallacies. There is one point worth addressing in the impotent arsenal of zombie arguments* presented by creationist dolts that has some relevance here. They claim endlessly that there are no transitional fossils, that there are no examples of one species changing into another, that a transitional eye is of zero use, and our vision is perfect as it is, and could not have come to be in incremental stages.

There are none so blind as those who will not see. In fact, the fossil record is replete with transitional forms. Any particular characteristic that one chooses to name has varied forms embedded in stone. Not only do we see eyes in myriad forms in fossils that show subtle, slight shifts toward our own versions, we see every stage in

* Unthinking and mindless, tired and drooling, relentlessly shuffling along, impervious to reason, intelligence, or debate, and desperately ugly.

living creatures too, from the photoreceptive patch in the simple, single-celled Euglena, all the way to us and the many creatures with sight far superior to our own. In fossils, the picture is far from complete, because fossils are rare and unlikely. But they are most definitely there, and we can with the most desultory glance see dozens of small moves from one to another—an arm to a wing, a photoreceptive cell to an eye, a fin to a foot, an oxygen-sucking spiracle to a lung. All these incremental changes over epochs are subject to selection, and all are encoded in DNA.

The truth is that as long as you reproduce, from a genomic point of view, you are transitional. Your DNA is a perfect transition between your parents' and your children's. If we had DNA from everyone who ever lived, and indeed every creature that ever lived, we could draw an impossibly colossal pedigree that charts every transition from cell to cell, parent to child, from species to species—every single colored pixel on an inconceivably gargantuan pageant of a color chart. We can't do that, unfortunately, so we use what genomic data we have, and compare living species and some dead to calculate the evolutionary distances traveled, and fit this new information in with the rest of the jigsaw: fossils, geology, statistics, mathematics.

It is in the accumulation of genetic change over time that new species are formed. Though the boundaries are smeary, as exemplified by the repeatedly violated boundary between *Homo sapiens* and *Homo neanderthalensis* (as argued in Chapter 1), over time, and with the unique pressures experienced by individual creatures, enough changes will be acquired and rooted in a population that they speciate. Their genes become so different that they are incompatible—the mechanics of sex, either physically or biochemically (or often both), are different enough that they are no longer capable of reproducing. This version of evolution is a real, testable fact of biology, seen in living species and in the dead. But conclusively, we have observed this in plenty of species in real time, since Darwin. The apple maggot fly is becoming two separate species as a result of generations feeding on different trees that fruit at different times. Since the introduction of apples in the United States in the early 1850s, some began feeding on apples instead of the hawthorn fruit. Now the two preferences have diverged to the extent that though they live in close proximity, apple feeders will not mate with hawthorn eaters. The European blackcap is a migrating bird, and most of them winter in Spain, but a few do so in the UK. Since the 1960s, when garden bird feeding became more popular, some blackcaps come to the UK, which is closer to their summer residence in Germany. They arrive before the Spanish migrants, and so get first dibs on mating, and now they are beginning to look like different species of bird.

All these birds and insects and we ourselves are transitional as we shift subtly and often cryptically over generational time. The creationists say there is real microevolution—the changes within a species—but there is no macroevolution—changes from one species to another. Biologists don't make the distinction. They are the same process over a long enough timescale, with the right conditions present. Unlike the blackcap, we, for now, are firmly within the microevolutionary mode. There is no prospect of us speciating; we're too similar, and too widespread, and we interbreed too much and too slowly. But over a long enough period, all species become something else living, or become dead. This is the continuous fact of life on Earth.

Darwin's theory of evolution by natural selection is just that—a theory. Outside science, there is some understandable confusion about what that word means. In science, a *theory* is the best description we have. Unlike the common usage, it's not a guess, or a hunch, or a hypothesis. It's the most complete subjective picture of the living world that we have. It's not truth, because that is the realm of math, religion, and philosophy. In science we simply lean in toward truth, every step inching us closer to the way things actually are, rather than how we perceive them to be, or would like them to be.

Darwinian evolution is a theory without peer. It doesn't have to compete with other theories, because it's the only game in town. There is no other scientific description of life on Earth that is supported by what we observe and what we test. Charles Darwin formulated his idea 50 years before genes, 100 before the double helix, and 150 before the human genome was read. But they all say the same thing. Life is a chemical reaction. Life is derived from what came before. Life is imperfect copying. Life is the accumulation and refinement of information embedded in DNA. Natural selection explains how, once it had started, life evolved on Earth. We busy ourselves refining the theory, and working out the details with a scrutiny and precision that has been enabled and invigorated by reading genome after genome, and crunching those numbers until comprehensible patterns emerge. We are the data.

Are humans still subject to an evolution by the forces of selection? Yes we are, though its grasp on us is weakened and slowed compared to every other species in our 4-billion-year family tree. We are animals: We are special. Are we still evolving? The answer is unequivocal. Evolution is change plus time. We've seen it in our deep and recent past, sometimes demonstrably the result of positive natural selection, and often just an unopposed drift through time. An unchanging species is already extinct. As long as we keep making new ones, human beings most beautiful and most wonderful have been, and are being, evolved.

Epilogue

> We shall not cease from exploration
> And the end of all our exploring
> Will be to arrive where we started
> And know the place for the first time.
>
> "Little Gidding," T. S. Eliot

Writing this book has affected the way I think about people. I write these words sitting on a crowded London Underground train on the Victoria Line on a Wednesday morning in a sullen March. The train is crammed, and it's not particularly pleasant; everyone is harassed about arriving at work on time, and we aren't prone to gracious behavior in these circumstances. I love people watching, though. I love looking at faces, at how they are so different, and so similar. This part of the journey is identical—we all got on at Brixton—but our 4-billion-year routes before this commute were all different. I look at my fellow commuters and marvel at our species. Yes, we are unique, and can do so many things that no other species can come close to. All species are unique, and we can't see in sixteen different wavelengths of light like the mantis shrimp does; we can't fly for a thousand miles without stopping to rest, like so many migrating birds; we can't breathe underwater.

All species are special. Our uniquely evolved intelligence has pushed us to be a technological species. We do all of those things because we invented science and engineering and culture, and strove to understand our world and ourselves. We are the species that looked in the mirror, but it wasn't vanity that prompted us to interrogate our own bodies and our evolution. It was curiosity. What a senseless phrase is "curiosity killed the cat." To be incurious is to be inhuman. We did look inward into the hidden kingdoms of anatomy, then cells, and now genes. We also looked up to the skies, and down into the ground and seas, and into the invisible worlds of the atoms, and subatoms, and now the quantum realm. We are the explorers, and science is exploration.

Any theory of biology has to incorporate the similarities and the differences between us all. We're special; we're just another ape. And so I look at the faces on the train. I see the lines on a woman's face and think of the proteins that give her skin that elasticity. And the spectrum of eyes, from the palest blue to as black as the night, or the scalloping of the backs of our front teeth, which I know without looking will be different in those Londoners with East Asian origins. I think about the genes that are conspiring in a community with all the others so that they will survive in this biological husk that they have built. The precise nature of the conspiracy is different in every one of these people. We're unique. Within our species we are all so very utterly unique too.

Unique is an odd word that needs or requests no adjectival qualifier, and on the first page I casually dismissed the uniqueness of our species. Nevertheless, there has never been anyone quite like you before, and there never will be again. Your face, your physiology, your metabolism, your experience, your family, your DNA, and your history are the contrivances of cosmic happenstance in a fully indifferent universe. We're unique in our DNA, but it was drawn from millions of past lives. It pleases me that there are people in this carriage descended from that dastard Richard III, or a man who gave Jesus' foreskin to a pope, and many more who are my close or distant cousins via some unknown European in the fifteenth century. Or a Viking who stepped onto the volcanic shore of Iceland for the first time. It pleases me that every one of us shares an ancestor who, in the words of Joseph Chang, sowed rice on the banks of the Yangtze or who labored to build the pyramids. It pleases me that our ancestors might have been the first to milk cows, or goats, or designed and made pots, or hunted boar, or mammoths, or even had sex with Neanderthals or Denisovan people, and in doing so brought them with us for as long as humans will endure.

In trying to understand who we are and how we came to be here now, we reconstruct the past. We're infinitely more than DNA, of course, and we've come a very long way. Using the length of this book as a timescale, recorded history is only equivalent to a single character, one letter among more than 685,000. But within that minuscule drop in the earth's oceans of time is all human history, recorded in stone, in writing, in art, by word, in bones, in buildings and kitchenware, and in DNA. The genome is a history book, and we will not cease from exploring it, and as long as there are people our exploring will never be at an end. Here, for now, is the end of this book.

ACKNOWLEDGMENTS

The following people have helped with words and ideas, and I am hugely grateful to all of them: Jennifer Raff, Ed Yong, Ewan Birney, Suzi Gage, Alex Garland, Siddhartha Muhkerjee, Kevin O'Byrne, Rasmus Nielson, Simon Fisher, Josh Akey, Ian Barnes, Andrew Millard, Stephen Keeler, Sarah Kent, Lady Anne Piper, Richard Dawkins, Peter Frankopan, Brian Cox, Aoife McLysaght, Louise Crane, Lara Cassidy, Yan Wong, Lee Rowen, Joseph Alberto Santiago, Nathaniel Comfort, Nathaniel Rutherford, Marcus Harben, Ana Paula Lloyd, Chris Gunter, Emma Darwin, Graham Coop, Lisa Matisoo-Smith, Jane Sowden, Elspeth Merry Price, David Rutherford, Ananda Rutherford, the Darwin Correspondence Project, Subhadra Das, Debbie Kennett, the Celeriacs, Razib Khan, Leonid Kruglyak, Alys Wardle, Andrew Cohen, Gail Kay, Henry Gee, Lena Kerans, Aylwyn Scally, Francesca Stavrakapoulou, Tamsin Edwards, Penny Young, Matt Ridley, Hamish Spencer, Kevin Mitchell, Marcus Munafo, Pete Etchells, Robert Plomin, Peter Donnelly, and Chris Stringer. At the BBC Radio Science Unit: Anna Buckley, Deborah Cohen, Sasha Feachem, Fiona Hill, Fiona Roberts, Adrian Washbourne, Andrew Luck-Baker, Jen Whyntie, and Marnie Chesterton, all of whom produced programs for me that were genetics-related. Tim Usborne and Paul Sen directed me in the BBC4 series *The Gene Code*, which influenced some of the ideas contained within this book.

Alice Roberts helped me more than could reasonably be expected, I think in atonement for the many times she has told me genetics is dull. Matthew Cobb's generosity with his help is only matched by his knowledge of Marvel comics. My literary agent Will Francis does nothing but make my writing better, so thank you to all at Janklow & Nesbit. The most profound gratitude to my editor Bea Hemming, whose editorial skill and grace are a joy to behold, to Holly Harley, Nicholas Cizek, and Charlotte Cole for their expert editing, and to Pete Garceau for the cover design. Thanks also to Dan O'Connor and Karen Wise, and to Sarah Schneider, Pamela Schechter, Jeanne Tao, Jennifer Hergenroeder, and the rest of the team at The Experiment. As ever, my deepest love and thanks go to Georgia, Beatrice, and Jake, and to the latest addition to my branch of human evolution, Juno.

GLOSSARY

allele: Although we all carry the same set of genes, they are not identical in all of us. An allele is a variant of a gene (or position in the genome) akin to an alternative spelling, or a typo. The US version of this book will contain the word *behavior*, whereas in the UK it is *behaviour*, which doesn't change its meaning, whereas the single letter change in *affect* and *effect* alters meaning from "to change" to "result." Both are examples of alleles.

amino acid: Small molecules that link together to form proteins. Each amino acid is encoded in DNA by a particular arrangement of three bases.

bases: Also called nucleotides, or referred to as letters; the individual components of DNA, the arrangement of which determines the genetic code. There are four bases, abbreviated to A, T, C, and G. The double helix in a sense is like a twisted ladder, the rungs of which are made of pairs of bases; A only ever pairs with T, C only ever with G. (In RNA, the T is replaced with a U.)

BCE: Before the Common Era. Synonymous with the term *Before Christ*, but science has replaced it, appropriately, with something universal.

CE: The Common Era, synonymous with the old-fashioned Anno Domini; e.g., the year this book was first published was 2016 CE.

chromosome: Long stretches of DNA that harbor genes. Species have a specific number of chromosomes. In humans it's twenty-two pairs of autosomes, plus two sex chromosomes: two

Xs if you are a woman, and an X and a Y if you are not.

codon: Three bases in a particular order encode a specific amino acid, which link together to form proteins. There are four bases, therefore sixty-four possible combinations if arranged into threes. But biology only uses twenty amino acids (plus a signal to end a protein, STOP), which means that the genetic code has redundancy in it: Several different codons can encode the same amino acid. A stretch of codons in DNA marks out a gene.

DNA: Deoxyribonucleic acid: the material of genetics, typically arranged into chromosomes. DNA is a script in which genes are written.

gene: In its simplest form, a gene is a sequence of DNA that encodes a working protein. This definition adheres to what is known as the "central dogma": DNA makes RNA makes protein. Genes are composed of DNA, and these lie as part of chromosomes, like sentences in text. There is no more precise definition, but there are active genetic elements that do not follow this model. In the last few years, many bits of DNA that encode RNA that never becomes a protein have been identified. These may yet be classified as genes.

genetics: The study of genes, DNA, disease, inheritance, evolution, and a whole lot more.

genome: The entire genetic material of an organism, the sum total of its DNA. The Human Genome Project was a huge publicly funded collaborative scientific endeavor that concluded in the first few years of the twenty-first century. Its aim—which was delivered under budget and on time—was to provide a complete readout of a human genome as a database that we could go on to mine for information about how humans work, how we evolved, and what happens when we go wrong. The principal output was a reference genome, an average human from which we can compare our infinite variation.

genomics: The study of genomes, including genetics but much more as well. In studying genomes we look at more than simply the genes that have specific functions in an organism, and analyze the regulation of those genes, and the information stored in DNA.

genotype: Genotype refers to the version of the genes that you possess.

heterozygous/homozygous: These refer to the pair of each gene you have, one inherited from each parent. Heterozygous means you have two different versions of the same gene; homozygous means they are the same.

Mendelian/Mendel's Laws: The basic rules of biological inheritance were set by Gregor Mendel in the nineteenth century, as a result of his experiments in which he bred thousands of pea plants and observed how various characteristics were passed from generation to generation. There are three broad rules, which can be summarized thus:

1. Organisms have two copies of each allele, one inherited from each parent.
2. Each trait is inherited independently of any other.
3. Some alleles are dominant over others, which are called recessive. If you inherit one dominant from one parent, and one recessive from the other, only the dominant will be expressed. To see a recessive trait, you must have inherited two recessive alleles, one from each parent—for example, with the alleles for red hair.

As is so often the case, these laws are more like rules, in that there are some notable exceptions. Biology is annoying like that.

mitochondria: Tiny compartments within cells that are primarily responsible for generating energy for that cell. They contain their own chromosome, a small loop of DNA separate from the vast majority contained within the nucleus of a cell. Mitochondrial DNA (which is sometimes written as mtDNA) is of great interest to geneticists and genealogists as it is only ever passed on from mother to child, and thus can exclusively chart a matrilineage.

phenotype: The phenotype is the physical manifestation of a gene or genotype. For example, a particular version of the gene MC1R (the genotype) will mean your phenotype will be a redhead.

protein: The primary functional biological molecules in living things. They are made up of simple molecules called amino acids assembled into long strings. These fold up to make a three-dimensional structure, and often assemble with other proteins in cells to enact their function. All life is made of, or by, proteins.

single nucleotide polymorphism: Many of the genetic differences between people are made up of individual letters of genetic code being different at specific points in the genome (see allele, above). When these are simply alterations in a single letter at a specific point, we call them single nucleotide polymorphisms, which is universally abbreviated to SNP (and pronounced "snip"). The words *inquiry* and *enquiry* would be an alphabetical equivalent of a SNP.

REFERENCES AND FURTHER READING

I urge you to read the following books that have influenced me over the years on the subjects of genetics, evolution, and humans.

The Language of the Genes: Solving the Mysteries of Our Genetic Past, Present and Future by Steve Jones (HarperCollins, 1991) is a classic, written well before the Human Genome Project was begun and so twenty-five years later is out of date. However, the stories within and Jones' peerless storytelling make it well worth reading to this day.

Genome: A Biography in 23 Chapters by Matt Ridley (Fourth Estate, 1999) is also a classic, and also published before the human genome was unveiled.

Life's Greatest Secret: The Race to Crack the Genetic Code by Matthew Cobb (Profile, 2015) is the most definitive version of the story of the genetic code that I have read.

And of course, always return to the source:

The Descent of Man, and Selection in Relation to Sex by Charles Darwin, MA, FRS (John Murray, 1871)

INTRODUCTION

Estimate of the total number of humans born in the last 50,000 years

Carl Haub, "How Many People Have Ever Lived on Earth?" Population Reference Bureau, prb.org/Publications/Articles/2002/HowManyPeopleHaveEverLivedonEarth.aspx

"There is no one alive more youer than you!" from *Happy Birthday to You!* by Dr. Seuss (HarperCollins Children's Books, 2005)

International Human Genome Sequencing Consortium, Lander, Eric S., et al., "Initial Sequencing and Analysis of the Human Genome," *Nature* 409 (2001), 860–921

CHAPTER 1: HORNY AND MOBILE

On the origin of life in hydrothermal vents

Many have written on the origin of life, including myself (*Creation*, Viking, 2013), but none better than Nick Lane, notably in *The Vital Question: Why Is Life the Way It Is?* (Profile, 2015)

The Hobbit

Brown, P., et al., "A New Small-Bodied Hominin from the Late Pleistocene of Flores, Indonesia," *Nature* 431 (2004), 1055–61

Sutikna, Thomas, et al., "Revised Stratigraphy and Chronology for *Homo floresiensis* at Liang Bua in Indonesia," *Nature* 532 (2016), 366–69

van den Bergh, Gerrit D., et al., "*Homo floresiensis*-like Fossils from the Early Middle Pleistocene of Flores," *Nature* 534 (2016), 245–48

On cryptozoology

Jobling, Mark A., "The Truth Is Out There," *Investigative Genetics* 4: 24 (2013)

On the Ebu Gogo

Wong, Kate, "The Littlest Human," *Scientific American* 16 (2006), 48–57

The first Neanderthal DNA

Krings, Matthias, et al., "Neandertal DNA Sequences and the Origin of Modern Humans," *Cell* 90: 1 (1997), 19–30

Forty-seven Chinese teeth

Liu, Wu, et al., "The Earliest Unequivocally Modern Humans in Southern China," *Nature* 526 (2015), 696–69

The stories of the Neanderthals

Howell, F. Clark, "The Evolutionary Significance of Variation and Varieties of 'Neanderthal' Man," *The Quarterly Review of Biology* 32: 4 (1957), 330–47

Rendu, William, et al., "Evidence Supporting an Intentional Neandertal Burial at La Chapelle-aux-Saints," *Science* 318: 5855 (2007), 1453

More Neanderthal genomes

Green, Richard E., et al., "Analysis of One Million Base Pairs of Neanderthal DNA," *Nature* 444: 7117 (2006), 330–36

Noonan, James P., et al., "Sequencing and Analysis of Neanderthal Genomic DNA," *Science* 314: 5802 (2006), 1113–18

Green R. E., Krause J., Briggs A. W., et al., "A Draft Sequence of the Neandertal Genome," *Science* 328: 5979 (2010), 710–22

Prüfer, Kay, et al., "The Complete Genome Sequence of a Neanderthal from the Altai Mountains," *Nature* 505 (2014), 43–49

On speech, FoxP2, people, and songbirds

Arensburg, B., et al., "A Middle Palaeolithic Human Hyoid Bone," *Nature* 338 (1989), 758–60

Murugan, Malavika, et al., "Diminished FoxP2 Levels Affect Dopaminergic Modulation of Corticostriatal Signaling Important to Song Variability," *Neuron* 80: 6 (2013), 1464–76

Krause, J., Lalueza-Fox, C., Orlando, L., et al., "The Derived FOXP2 Variant of Modern Humans Was Shared with Neandertals," *Current Biology* 17: 21 (2007), 1908–12

Hurst, J. A., et al., "An Extended Family with a Dominantly Inherited Speech Disorder," *Developmental Medicine Child Neurology* 32: 4 (1990), 352–25

Kuhlwilm, M., et al., "Ancient Gene Flow from Early Modern Humans into Eastern Neanderthals," *Nature* 530 (2016), 429–33

On smelling, skin, and red hair

Mainland, Joel D., et al., "The Missense of Smell: Functional Variability in the Human Odorant Receptor Repertoire," *Nature Neuroscience* 17 (2014), 114

Lalueza-Fox, C., et al., "A Melanocortin 1 Receptor Allele Suggests Varying Pigmentation Among Neanderthals," *Science* 318: 5855 (2007), 1453–55

Hoover, Kara C., et al., "Global Survey of Variation in a Human Olfactory Receptor Gene Reveals Signatures of Non-Neutral Evolution," *Chemical Senses* 40 (2015), 481–88

On the impact of admixture with Neanderthals

Juric, Ivan, Aeschbacher, Simon, and Coop, Graham, "The Strength of Selection Against Neanderthal Introgression," *bioRxiv* (October 30, 2015)

Harris, Kelley, and Nielsen, Rasmus, "The Genetic Cost of Neanderthal Introgression," *bioRxiv* (March 29, 2016)

Enter Denisova

Krause, J., Fu, Q., Good, J. M., et al., "The Complete Mitochondrial DNA Genome of an Unknown Hominin from Southern Siberia," *Nature* 464 (2010), 894–97

Sawyer, Susanna, et al., "Nuclear and Mitochondrial DNA Sequences from Two Denisovan Individuals," *PNAS* 112 (2015), 15696–700

Reich, D., Richard, E. G., et al., "Genetic History of an Archaic Hominin Group from Denisova Cave in Siberia," *Nature* 468 (2010), 1053–60

Reich, David, et al., "Denisova Admixture and the First Modern Human Dispersals into Southeast Asia and Oceania," *American Journal of Human Genetics* 89 (2011), 516–28

Huerta-Sanchez, Emilia, et al., "Altitude Adaptation in Tibetans Caused by Introgression of Denisovan-Like DNA," *Nature* 512 (2014), 194–97

Curnoe, D., et al., "A Hominin Femur with Archaic Affinities from the Late Pleistocene of Southwest China," *PLOS One* 10: 12 (2015)

Birney, Ewan, and Pritchard, Jonathan K., "Archaic Humans: Four Makes a Party," *Nature* 505 (2014), 32–34

CHAPTER 2: THE FIRST EUROPEAN UNION

Hardy, Karen, et al., "The Importance of Dietary Carbohydrate in Human Evolution," *The Quarterly Review of Biology* 90: 3 (2015), 251

Itan, Y., et al., "The Origins of Lactase Persistence in Europe," *PLOS Computational Biology* 5: 8 (2009)

Shennan, Stephen, et al., "Regional Population Collapse Followed Initial Agriculture Booms in Mid-Holocene Europe," *Nature Communications* 4 (2013)

Lazaridis, Iosif, et al., "Ancient Human Genomes Suggest Three Ancestral Populations for Present-Day Europeans," *Nature* 513 (2014), 409–13

Seguin-Orlando, Andaine, et al., "Genomic Structure in Europeans Dating Back at Least 36,200 Years," *Science* 346: 6213 (2014)

Fu, Qiaomei, et al., "Genome Sequence of a 45,000-Year-Old Modern Human from Western Siberia," *Nature* 514 (2014), 445–50

Helgason, et al., "Sequences from First Settlers Reveal Rapid Evolution in Icelandic mtDNA Pool," *PLOS Genetics* 5: 1 (2009)

Benedictow, O. J., *The Black Death 1346–1353: The Complete History* (Boydell Press, 2004)

Procopius, *Secret History: History of the Wars*, II. xxii–xxxiii (translated by Richard Atwater. Chicago: P. Covici, 1927; New York: Covici Friede, 1927; reprinted Ann Arbor, MI: University of Michigan Press, 1961)

Adhikari, K., et al., "A Genome-Wide Association Scan in Admixed Latin Americans Identifies Loci Influencing Facial and Scalp Hair Features," *Nature Communications* 7 (2016)

Leslie, Stephen, et al., "The Fine-Scale Genetic Structure of the British Population," *Nature* 519 (2004), 309

CHAPTER 3: THESE AMERICAN LANDS

The Saga of Erik the Red, Sephton, J. (trans.), *Icelandic Saga Database*, Sveinbjorn Thordarson (ed.), sagadb.org/eiriks_saga_rauda.en

Rasmussen, Morten, "The Ancestry and Affiliations of Kennewick Man," *Nature* 523 (2015), 455–58

Zimmer, Carl, "New DNA Results Show Kennewick Man Was Native American," *The New York Times* (June 18, 2015)

Mulligan, Connie J., and Em?ke J. E. Szathmáry, "The Peopling of the Americas and the Origin of the Beringian Occupation Model," *American Journal of Physical Anthropology* 162: 3 (2017), 403–408

Amorim, Carlos Eduardo G., et al., "Genetic Signature of Natural Selection in First Americans," *PNAS* 114: 9 (2017), 2195–99

Fumagalli, Matteo, et al., "Greenlandic Inuit Show Genetic Signatures of Diet and Climate Adaptation," *Science* 349: 6254 (2015), 1343–47

Han, Eunjung, et al., "Clustering of 770,000 Genomes Reveals Post-Colonial Population Structure of North America, *Nature Communications* 8: 14238 (2017)

Bryc, Katarzyna, et al., "The Genetic Ancestry of African Americans, Latinos, and European Americans Across the United States," *American Journal of Human Genetics* 96: 1 (2015), 37–53

Harmon, Amy, "Indian Tribe Wins Fight to Limit Research of Its DNA," *The New York Times* (April 21, 2010)

"After Havasupai Litigation, Native Americans Wary of Genetic Research," *American Journal of Medical Genetics* 152: 7 (2010), ix

CHAPTER 4: WHEN WE WERE KINGS

Charlemagne

Rohde, Douglas L. T., Olson, Steve, and Chang, Joseph T., "Modelling the Recent Common Ancestry of All Living Humans," *Nature* 431 (2004), 562–66

Chang, Joseph, "Recent Common Ancestors of All Present-Day Individuals," *Advances in Applied Probability* 31 (1999), 1002–26

Ralph, Peter, and Coop, Graham, "The Geography of Recent Genetic Ancestry Across Europe," *PLOS Biology* 11: 5 (2013)

"Revealed: The Indian Ancestry of William," *Times* (June 14, 2013)

Lucotte, Gérard, et al., "Haplogroup of the Y Chromosome of Napoléon the First," *Journal of Molecular Biology Research* 1: 1 (2011)

Richard III

Seguin-Orlando, Andaine, et al., "Identification of the Remains of King Richard III," *Nature Communications* 5 (2014)

Edwards, Russell, *Naming Jack the Ripper: New Crime Scene Evidence, A Stunning Forensic Breakthrough, The Killer Revealed* (Sidgwick & Jackson, 2014)

The collapse of the Hapsburgs

Alvarez, Gonzalo, Ceballos, Francisco C., and Quinteiro, Celsa, "The Role of Inbreeding in the Extinction of a European Royal Dynasty," *PLOS ONE* 4: 4 (2009)

Álvarez, Gonzalo, Ceballos, Francisco C., and Berra, Tim M., "Darwin Was Right: Inbreeding Depression on Male Fertility in the Darwin Family," *Biological Journal of the Linnean Society* 114 (2015), 474–83

Fareed, Mohd, and Afzal, Mohammad, "Estimating the Inbreeding Depression on Cognitive Behavior: A Population Based Study of Child Cohort," *PLOS ONE* 9: 10 (2014)

McQuillan, Ruth, et al., "Evidence of Inbreeding Depression on Human Height," *PLOS GENETICS* 8: 7 (2012)

Mendizabal, Isabel, et al., "Reconstructing the Population History of European Romani from Genome-Wide Data," *Current Biology* 22: 24 (2012), 2342–49

Kalaydjieva, Luba, et al., "Genetic Studies of the Roma (Gypsies): A Review," *BMC Medical Genetics* 2 (2001), 5

Waller, John C., et al., "Prevalence of Congenital Anomaly Syndromes in a Spanish Gypsy Population," *Journal of Medical Genetics* 29: 7 (1992), 483

Gazal, Steven, et al., "High Level of Inbreeding in Final Phase of 1000 Genomes Project," *Scientific Reports* 5 (2015)

CHAPTER 5: THE END OF RACE

Uglow, Jenny, *The Lunar Men: The Friends Who Made the Future 1730–1810* (Faber and Faber, 2003)

Galton on cakes

Galton, Francis, "Cutting a Round Cake on Scientific Principles" *Nature* 75 (1906), 173

Galton, Francis, "On the Anthropometric Laboratory at the Late International Health Exhibition," *Journal of the Anthropological Institute* 14 (1884), 12

Galton, Francis, *Hereditary Genius: An Inquiry into Its Laws and Consequences* (Macmillan, 1869)

Hirschfeld, Ludwik, and Hirschfeld, Hanka, "Serological Differences Between the Blood of Different Races: The Result of Researches on the Macedonian Front," *The Lancet* 194: 5016 (1919), 673–718

Schneider, W. H., "The History of Research on Blood Group Genetics: Initial Discovery and Diffusion," *History and Philosophy of the Life Sciences* 18 (1996), 282

On Rosalind Franklin's grandfather

Piper, Anne, "Light on a Dark Lady," *Trends in Biochemical Sciences*, 23 (1998), 151–54, cwp.library.ucla.edu/articles/franklin/piper.html

Richard Lewontin's classic study on the biology of race

Lewontin, R. C., "The Apportionment of Human Diversity," *Evolutionary Biology* 6 (1972), 381–98

This paper has been subject to much scrutiny over the years, much worth reading, as it is not a simple subject. Richard Dawkins and Yan Wong discuss it in *The Ancestor's Tale: A Pilgrimage to the Dawn of Evolution* (Weidenfeld & Nicolson, 2005), and Anthony Edwards critiqued it in 2003 in a paper entitled "Human Genetic Diversity: Lewontin's Fallacy" (*BioEssays* 25: 8, 798–801).

On earwax, hair thickness, and tooth shoveling

Sato, T., et al., "Allele Frequencies of the ABCC11 Gene for Earwax Phenotypes Among Ancient Populations of Hokkaido, Japan," *Journal of Human Genetics* 54: 7 (2009), 409–13

Nakano, Motoi, et al., "A Strong Association of Axillary Osmidrosis with the Wet Earwax Type Determined by Genotyping of the ABCC11 Gene," *BMC Genetics* 10: 42 (2009)

Kamberov, Y. G., et al., "Modeling Recent Human Evolution in Mice by Expression of a Selected EDAR Variant," *Cell* 152 (2013), 691–702

Noah Rosenberg's important study of race and DNA

Rosenberg, N. A., et al., "Genetic Structure of Human Populations," *Science* 298: 5602 (2002), 2381–85

Raff, Jennifer, "Nicholas Wade and Race: Building a Scientific Façade" *Violent Metaphors* (blog) (May 21, 2014), violentmetaphors.com/2014/05/21/nicholas-wade-and-race-building-a-scientific-facade

Reuter, Shelley, "The Genuine Jewish Type: Racial Ideology and Anti-Immigrationism in Early Medical Writing about Tay-Sachs Disease," *The Canadian Journal of Sociology* 31: 3 (2006), 291–323

Hughey, Matthew W., and Goss, Devon R., "A Level Playing Field? Media Constructions of Athletics, Genetics, and Race," *ANNALS of the American Academy of Political and Social Science* 661: 1 (2015), 182–211

Vancini, R. L., et al., "Genetic Aspects of Athletic Performance: The African Runners Phenomenon," *Open Access Journal of Sports Medicine* 5: (2014), 123–27

CHAPTER 6: THE MOST WONDROUS MAP EVER PRODUCED BY HUMANKIND

ENCODE Project Consortium, Birney, E., et al., "Identification and Analysis of Functional Elements in 1% of the Human Genome by the ENCODE Pilot Project," *Nature* 447: 7146 (2007), 799–816

On the origins of hemophilia A, and its description in the Talmud

Rosner, Fred, and Pierce, Glenn F., "Correspondence: Hemophilia A," *New England Journal of Medicine* 330: 1617 (1994)

Waller, John C., "The Birth of the Twin Study—A Commentary on Francis Galton's 'The History of Twins,'" *International Journal of Epidemiology* 41: 4 (2012), 913–17

Martinez-Frias, M. L., and Bermejo, E., "Prevalence of Congenital Anomaly Syndromes in a Spanish Gypsy population," *Journal of Medical Genetics* 29 (1992), 483–86

Vanscoy, L. L., et al., "Heritability of Lung Disease Severity in Cystic Fibrosis," *American Journal of Respiratory and Critical Care Medicine* 175 (2007), 1036

Manolio, Teri A., et al., "Finding the Missing Heritability of Complex Diseases," *Nature* 461: 7265 (2009), 747

The birth of GWAS

Klein, R. J., et al., "Complement Factor H Polymorphism in Age-Related Macular Degeneration," *Science* 308: 5720 (2005), 385–89

The Wellcome Trust Case Control Consortium, "Genome-Wide Association Study of 14,000 Cases of Seven Common Diseases and 3,000 Shared Controls," *Nature* 447 (2007), 661–78

Sturm, R. A., and Larsson, M., "Genetics of Human Iris Color and Patterns," *Pigment Cell & Melanoma Research* 5 (2009), 544

Physics joke: the Alpher-Bethe-Gamow paper

Alpher, R. A., Bethe, H., and Gamow, G., "The Origin of Chemical Elements," *Physical Review* 73: 7 (April 1, 1948), 803–4

Tongue rolling

Matlock, P., "Identical Twins Discordant in Tongue-Rolling," *Journal of Heredity* 43 (1952), 24

Sturtevant, A. H., "A New Inherited Character in Man," *Proceedings of the National Academy of Sciences USA* 26 (1940), 100–2

Latham, Jonathan, "The Failure of the Genome," *Guardian* (April 17, 2011)

James, Oliver, "Sorry, but You Can't Blame Your Children's Genes," *Guardian* (March 30, 2016)

CHAPTER 7: FATE

State of Tennessee v. Davis Bradley Waldroup, Jr. (2011) Criminal Court for Polk County No. 08-101

Brooks-Crozier, Jennifer, "The Nature and Nurture of Violence: Early Intervention Services for the Families of MAOA-LOW Children as a Means to Reduce Violent Crime and the Costs of Violent Crime," *Connecticut Law Review* 44: 2 (2011)

Lenders, J. W. M., et al., "Specific Genetic Deficiencies of the A and B Isoenzymes of Monoamine Oxidase Are Characterized by Distinct Neurochemical and Clinical Phenotypes," *Journal of Clinical Investigation* 97: 4 (1996), 1010–19

Frazzetto, G., et al., "Early Trauma and Increased Risk for Physical Aggression During Adulthood: The Moderating Role of MAOA Genotype," *PLOS ONE* 2: 5 (2007)

Gibbons, Ann, "Tracking the Evolutionary History of a 'Warrior' Gene," *Science* 304: 5672 (2004), 818

Caspi, A., et al., "Role of Genotype in the Cycle of Violence in Maltreated Children," *Science* 297: 5582 (2002), 851–54

Lea, Rod, and Chambers, Geoffrey, "Monoamine Oxidase, Addiction, and the 'Warrior' Gene Hypothesis," *New Zealand Medical Journal* 120: 1250 (2007)

McDermott, Rose, et al., "Monoamine Oxidase A gene (MAOA) Predicts Behavioral Aggression Following Provocation," *PNAS* 106: 7 (2009), 2118–23

Beaver, Kevin M., et al., "Monoamine Oxidase A Genotype Is Associated with Gang Membership and Weapon Use," *Comprehensive Psychiatry* 51: 2 (2010), 130–34

"'Ruthlessness Gene' Discovered by Michael Hopkin," *Nature* (April 4, 2008)

Hunter, Philip, "The Psycho Gene," *EMBO Reports* 11: 9 (2010), 667–69

Tiihonen, J., et al., "Genetic Background of Extreme Violent Behavior," *Journal of Molecular Psychiatry* 20: 6 (2015), 786–92

Hogenboom, Melissa, "Two Genes Linked with Violent Crime," BBC Online (October 28, 2014)

On Adam Lanza

Kolata, Gina, "Seeking Answers in Genome of Gunman," *The New York Times* (December 24, 2012)

Etchells, Peter J., et al., "Prospective Investigation of Video Game Use in Children and Subsequent Conduct Disorder and Depression Using Data from the Avon Longitudinal Study of Parents and Children," *PLOS ONE* 11: 1 (2016)

Myers, P. Z., "Fishing for Meaning in a Dictionary of Genes," *Pharyngula* (December 27, 2012)

The Hongerwinter

Banning, C., "Food Shortage and Public Health, First Half of 1945," *The Annals of the American Academy of Political and Social Science* 245: The Netherlands During German Occupation (May 1946), 93–110

Stein, A. D., and Lumey, L. H., "The Relationship Between Maternal and Offspring Birth Weights After Maternal Prenatal Famine Exposure: The Dutch Famine Birth Cohort Study," *American Journal of Human Biology* 72: 4 (2000), 641–54

Kaati, G., et al., "Cardiovascular and Diabetes Mortality Determined by Nutrition During Parents' and Grandparents' Slow Growth Period," *European Journal of Human Genetics* 10: 11 (2002), 682–88

Pembrey, Marcus, et al., "Human Transgenerational Responses to Early-Life Experience: Potential Impact on Development, Health and Biomedical Research," *Journal of Medical Genetics* 51: 9 (2014), 563–72

Chopra, Deepak, and Tanzi, Rudolph, "You Can Transform Your Own Biology," chopra.com/ccl/you-can-transform-you r-own-biology

CHAPTER 8: A SHORT INTRODUCTION TO THE FUTURE OF HUMANKIND

Fu, W., O'Connor, T. D., et al., "Analysis of 6,515 Exomes Reveals the Recent Origin of Most Human Protein-Coding Variants," *Nature* 493 (2013), 216–20

Pandit, Jaideep J., et al., "A Hypothesis to Explain the High Prevalence of Pseudo-Cholinesterase Deficiency in Specific Population Groups," *European Journal of Anaesthesiology* 28 (2011), 550

Reich, D., et al., "Reconstructing Indian Population History," *Nature* 461 (2009), 489–94

Moorjani, Priya, et al., "Genetic Evidence for Recent Population Mixture in India," *The American Journal of Human Genetics* 93: 3 (2013), 422–38

Bolund, Elisabeth, et al., "Effects of the Demographic Transition on the Genetic Variances and Covariances of Human Life-History Traits," *Evolution* 69 (2015), 747–55

Rolshausen, Gregor, et al., "Contemporary Evolution of Reproductive Isolation and Phenotypic Divergence in Sympatry Along a Migratory Divide," *Current Biology* 19: 24 (2009), 2097–101

TEXT AND IMAGE CREDITS

page 6, excerpt from "This Be the Verse" from *The Complete Poems of Philip Larkin* by Philip Larkin, edited by Archie Burnett.

page 20, reproduced with permission from Chris Stringer

pages 38–39, John Gilkes

page 99, reproduced and redrawn by Stephen Leslie

page 225, taken from Galton, Francis, "On the Anthropometric Laboratory at the International Health Exhibition," *Journal of the Anthropological Institute* 14 (1884), 12 and reproduced with permission from Wiley

page 270, courtesy of Cold Spring Harbor Laboratory.

page 361, excerpt from Eliot, T. S., "Little Gidding" from *Four Quartets* (Faber and Faber, 1942). Reproduced with permission from Faber and Faber Ltd.

INDEX

Note: Page numbers in *italics* indicate figures.

23andMe, 50, 152–53, 155, 167–70, 175, 224, 305
314.1C and 315.1C mutations, 192
370A mutations, 244–47
1000 Genomes project, 211–12

A

ABCC11 genes, 239, 243
abuse, of children, 315, 319
Accu-Metrics, 145–47
Acton, Thomas, 208
adaptationism, 237, 243, 260, 343
ADHD (attention deficit hyperactivity disorder), 318
admixture introgression, 51–54, 58–59, 103, 139–40
The Adventures of Dollie, 208
Africa
 Galton's views on, 226–27
 genetic variation in, 51, 218, 240, 244, 248, 258–59
 inbreeding and, 152, 212
 infant mortality, 354
 migration from, 8, 36, *38–39*, 49, 52, 124, 260–61
 milk drinking, 75–76
 as origin of *Homo sapiens*, 1–2, 36, *38–39*, 56–57, 73–74
 paucity of ancient DNA from, 125, 151–52
 as racial group, 248–52, 264
 skin color in, 82–83, 258, 264
African Americans, 131, 152, 214, 217, 257–58. *See also* black people
aggression, 313–18
agriculture. *See* farming
Akey, Josh, 348
albinism, oculocutaneous, 210
alcoholism, 148, 255, 318
Allatius, Leo, 158n
alleles
 in blood groups, 234–35
 defined, 199, 366, 368
 dominant, 239, 305–7, 368
 for earwax, 239–40
 recessive, 200, 305–7, 368
 for red hair, 89–93, 169
 See also genes
alpha-actinin-3 genes, 259
Altai Mountains, 51, 55–56, 84
alternative splicing, 290
Althing, 106–7
altitude, adaptations to, 58–59, 258, 264
Alvarez, Gonzalo, 200–202, 204
Alzheimer's disease, 110, 168, 313
American Society for Human Genetics, 150–51
Americas
 European discovery, 127, 128–30
 indigenous population, 130–36, 139–40, 142–44, 146, 149–50
 North America, 74, 106, 133, 136, 141, 151–52

remains found in, 136–39
South America, 130, 134–35, 138, 164–65
See also Native Americans; United States
amino acids, 87, 239, 366, 367, 369
amygdala, 318
amylase genes, 69, 73
ancestors, common, 37, 57, 66, 122, 161, 164
AncestryDNA, 151
ancestry, testing for, 50, 90, 145–47, 151–56, 167–77, 224, 305
Anders, Bill, 266
androstenone, smell of, 46–47
anesthetics, 349–51
Angles, 96, 103–4
Anglia Transwalliana, 100–101
Anglo-Saxons, 102–3, 226
animals, domestication of, 68–69, 80, 340
Anna of Hapsburg, 197, *198*
Anne of Green Gables (Montgomery), 86
Anne of York, 185
Anthropomorphic Laboratory, 224–26
anxiety, 318, 327
Anzick, 136–39, 141
apes
to ape-men, image of, 15–16, 18–19
bipedalism in, 21
cryptids of, 26n
great apes (Hominidae), 22–23, 43, 59, 62
apple maggot flies, 359
Arabs, Galton's views on, 227, 249
archaea, 34
Arched-Frame Portrait, 179, 186
Ashkenazi Jews, 256–57
Asians
eastern, 67–68, 165, 235–36, 240, 248–52
southern, 130, 212
astrology, genetic, 145, 166, 176–77
athletics, 257–60
attention deficit hyperactivity disorder (ADHD), 318
Australia, 74, 82, 249, 251
Australopithecus afarensis, 21–22
autism spectrum disorder, 301, 313, 321
autosomes, 33–35, 145, 366
Avon Longitudinal Study, 335
axillary osmidrosis, 243

B

babies
crying in, 237
mortality of, 118, 201, 204–5, 353–56
See also children
Backer-Dirks family, 158–59
bacteria, 34, 39, 113–17, 121–24
baldness, 92n
barnacles, 16n, 222, 254
bases, nucleotide, 27, 29, 366, 367, 369
basketball, 260
bats, 341–42
Battle of Bosworth, 178, 180, 185n
Bayout, Abdelmalek, 315–16, 319
BCE (Before the Common Era), defined, 366
Beaker people, 94
beards, red hair in, 92–93
Beothuk, 146–47
Beringian Standstill, 135, 141
Bering Strait, 75, 131–32, 164, 245
Berryman, Michael, 243n–44n
Betteridge's Law, 328n
betting book, for gene sweepstakes, 269–72
Beveridge, William, 229–30
Bible, 171, 287–88. *See also* Christianity
biological evolution. *See* evolution
biometrics, *225–26*, 232–34
bipedalism, 21
bipolar disorder, 295, 313
birds, 45, 227–28, 341–42, 359

Birney, Ewan, 268–72, 278
Björk, 112
blackcaps, 359
Black Death, 116, 118–20
black people, 152, 226, 234n–235, 238, 248–50, 257–60, 263–64. *See also* African Americans
blood groups, 234–35
Blood Quantum, 144
Bluefish Caves, 133, 135
Blumenbach, Johann, 249
boar, smell of, 46–47
Bodmer, Walter, 97–98
Bolund, Elisabeth, 355–56
bones, DNA extracted from, 37–42, 56–57, 79
Boveri, Theodor, 304
Boxgrove Man, 95
Boyce, Carol Reynolds, 146–47
Boyd, James, 149
Bradshaw, Leslie, 311–12
brain capacity, 36
brain scans, 323–24
breast size, 246
Britain. *See* British Isles; England; Great Britain
BritainsDNA, 167, 169, 171–75
British Isles
  genetic map of, *99*
  People of the British Isles project, 97–101, 104–5, 108, 152
  See also England; Great Britain
Broca's area, 43
Brunner, Han, 313–14
bubonic plague, 115–17, 119, 121
Bulgaria, 66, 209
burials, 118, 136–38, 181–83, 194
butyrylcholinesterase (BCHE), 350–51

C

Call of Duty games, 320
Campbell, Eddie, 189n
cancer, 267, 290–91, 308, 313, 332, 337–38
Carandini family, 159
Cartman, Eric, 91n
Cassidy, Lara, 51n
caste system, 349–52
Catawba, 154–56
cats, 345
CDH13 genes, 317–18
CE (Common Era), defined, 366
celiac disease, 84
Celts, 101–2
central dogma, 277, 367
centromeres, 59
cerumen, 239
CFTR genes, 291–92
Chang, Joseph, 161–65, 363
Charlemagne, 157–60, 162, 166, 199
Charles II of Austria, 197, *198*
Charles II of Spain, 195–202
cheese, 78
Cherokee, 132n, 146, 148
children
  abuse of, 315, 319
  of the Hongerwinter, 331–32, 334–35
  mortality of, 118, 201, 204–5, 353–56
  number per family, 353–55
  suspicions of kidnapping of, 208–9
chimpanzees, 45, 48–49, 59, 69
China, 36, 58, 61, 67–68, 227, 245–46, 249
chips, gene, 296
Christianity, 3, 106n, 109, 287–88, 357. *See also* Jesus Christ
chromatography, 78
chromosomes
  damaged, 44, 211, 267, 292
  defined, 366–67
  of Denisovans and Neanderthals, 59
  of Hominidae, 59
  multiple copies of, 267, 273n
  sex, 33–35, 183, 304, 366–67
  X, 33–34, 54, 299n, 302, 313, 333, 345–46, 367

Y, 5, 33–35, 54, 57–58, 107–8, 173–75, 183–85, 367
See also DNA (deoxyribonucleic acid)
Churchill, Winston, 229
CHX10 genes, 275, 281, 283, 285–86
circumcision, 302
classification, 21–24
climate change, 88–90, 171, 347
Clinton, Bill, 285, 290
Clovis, 134–36
Cobb, Matthew, 47, 284n
codons, defined, 367
Cohen (surname), 216n
Cohen, Jacob, 231n
Cold Spring Harbor, 265–66, 268–69, 271
Collins, Francis, 285, 289
color
eyes, 168–69, 186, 208, 210, 304, 305–7
hair, 47–48, 82–83, 86–93, 169, 210, 298, 303
perception of, 225, 344–46
skin, 74, 81–85, 89, 208–10, 248–50, 257–60, 264
color blindness, 345–46
Columbus, Christopher, 127, 129, 133
Colville tribes, 138, 149
Combe, George, 329
combined pituitary hormone deficiency, 201
comics, superheroes in, 339, 342n–43n
cones, retinal, 275, 344–45
consanguinity, 205–8, 210–12
Constantinople, 116, 119
consumer genetics, 50, 90, 145–47, 151–56, 167–77, 224, 305
Conti, Tom, 172–73
convergence, 341
cooking, 69, 340
Coon, Carleton, 249
Coop, Graham, 53, 163
copy number variations (CNVs), 301
cousins, marriage between, 199–200, 203–4, 206–7, 210–11
Creation (Rutherford), 253
creationism, 95n, 337, 345n, 357, 359
creeps *vs.* jerks, 15
Crick, Francis, 16, 231n, 233, 277–78, 283–84
Crick, Walter Drawbridge, 231n
criminal behavior, 187–92, 311–22, 323–24, 329
Cromwell, Dean, 257
crying, in babies, 237
cryptids, 26n
Cumberbatch, Benedict, 181, 194
curiosity, as catalyst, 362
cystic fibrosis, 90–91, 291–94

D

Daily Mail, 88, 327
Daily Telegraph, 172–73
dairy farming, 78–81, 237
Danish people, 103–4
Dark Ages, 102–3
Darwin, Charles
The Descent of Man, 17, 262–63, 357
at Down House, 18, 232
family of, 204–5, 210, 221, 227
Galton and, 221–23, 227–28, 232–33
health of, 203–4, 220
on natural selection, 219n–20, 336–37, 359–60
On the Origin of Species, 1, 3–4, 16, 220, 254, 263
on race, 214, 223, 262–63
travels of, 129, 220, 254
Darwin, Emma (née Wedgwood), 203–5, 223n, 228
Darwin, Erasmus, 221, 227
Darwin, Leonard, 203–4, 222n
Darwin, Robert, 203n
Dawkins, Richard, 255
Dawson, Charles, 95
Dead Famous DNA, 324–25
deCODE, 110–11

Denisovans
chromosomes of, 59, 140
discovery of, 55–58, 60, 64n
interbreeding by, *20*, 60–61, 63, 261
taxonomic status, 61–62
Denmark, 80, 102, 108, 172
Denny, Paul, 271
deoxyribonucleic acid. *See* DNA (deoxyribonucleic acid)
The Descent of Man (Darwin), 17, 262–63, 357
diabetes, 143, 295
diet, 69–70, 80–81, 84–85, 139–41, 350–51
dinosaurs, 40, 341
dirt, mtDNA extracted from, 64n
diseases
Bill Clinton on, 290
eradication of, 90–91, 257, 291, 340
genetic risks for, 168, 207–8, 212, 292, 297, 300–302
race and, 237–38, 256–57
recessive, 111, 200–202, 206–8, 210–12, 256, 291
See also specific diseases
distal renal tubular acidosis, 201
dizygotic twins, 87, 297
DNA (deoxyribonucleic acid)
commercial testing for, 50, 90, 145–47, 152–56, 167–77, 224, 305
defined, 367
double helix, 16, 27, 33, 233, 274, 366
extracted from bones, 37–42, 56–57, 79
extracted from dirt, 64n
extracted from hair, 173, 325–26
extracted from teeth, 25, 72–73, 79, 108, 117, 121, 146
as historical source, 4, 8, 98, 150
junk DNA, 277–78
mitochondrial (*see* mitochondrial DNA [mtDNA])
Neanderthal, 37–40, 42–45, 47–50, 51–54, 88
non-coding, 277–79
replication, 347
sequencing, 7, 27–31, 41, 50, 162–63, 284, 321
See also chromosomes; genes; genomes
DNA Consultants, 146
Doggerland, 94, 96
dogs, 68, 202, 276
domains, in classification, 22
domestication of animals, 68–69, 80, 340
dominant alleles, 239, 305–7, 368
Donnelly, Peter, 97
double helix, 16, 27, 33, 233, 274, 366
Down House, 18, 232
Down syndrome, 236n, 267
drosophila, 32, 120n–21n
Duldig, Wendy, 186
Dutch people, 158–59, 260, 313–14, 330–35

E

earwax, 168, 239–40, 242–43
East Asians, 67–68, 165, 235–36, 240, 248–52
East Smithfield Black Death cemetery, 118
Ebu Gogo, 26n
EDAR genes, 243–46, 250
Eddowes, Catherine, 189–92
Edmonstone, John, 223
Edward III, King, 184–85
Edward IV, King, 179
Edward V, King, 179–80
Edwards, Russell, 189–93
eggs (human), 4–6, 34, 52, 297
Egypt, 116–17
Einhard, 157
Elizabeth I, Queen, 303
E-M34, 173
EMBO Reports, 316–17
endogamy, 209, 251, 350–52
England, 103–5, 169, 178, 180–81, 184, 205, 230. *See also* Great Britain

EPAS1 genes, 58–59
epicanthic folds, 235–36, 248
epigenetics, 332–35, 337–38
Ericson, Leif, 127
Ethiopia, 125n, 173, 249, 258
eugenics, 9, 148, 203–4, 223, 229–30, 257, 323
eukaryotes, 34
eumelanin, 82–83, 86–87
Europe
early *Homo* inhabitants, 66–67, 72–74
impact of farming and cooking, 68–70
infant mortality, 353–54
milk drinking, 75–81
skin color in, 82–86
See also specific countries
Evershed, Richard, 78
evolution
ape-man to man-ape, image of, 15–16, 18–19
creationism and, 95n, 337, 345n, 357, 359
current and future, in humans, 81, 344, 348, 352–56, 360
Darwin's views on, 16n, 17, 220, 222, 336
Lamarck's views on, 336–37
microevolution *vs.* macroevolution, 359
religion and, 59–60
role of mutations in, 44n, 347–48
as shrub instead of tree, 18–*20*
as theory without peer, 359–60
See also mutations; natural selection
exomes, 274
exons, 281
eyes
color, 168–69, 186, 208, 210, 304, 305–7
fossil record, 357–58
vision, 225, 344–46

F

facial hair, 92–93
false paternity, 185
Fame, 214, 218
famine, effects of, 330–32, 334–35, 337
farming, 68–75, 80–81, 84–85, 102, 126, 238, 347–49. *See also* dairy farming
fast twitch muscle fibers, 258
fate, genes as, 10, 42, 48, 168, 305, 337–38
fatty acid desaturases (FADS), 140
fertilization (human), 4–5
fingerprinting, DNA, 184, 192
Finland, 212, 258n, 317–18, 355
first cousin marriages, 199–200, 203–4, 206–7, 210–11
Fisher, Ronald, 204
Fisher, Simon, 44
fitness, reproductive, 200n
Fitzroy, Robert, 129
fleas, 113–15, 117, 122–24, 273
flies, 32, 120n–21n, 304–5, 359
flight
bat and bird, 341–42
human, 339–40, 343
Flores, 24–26, 62
Forer effect, 176
fossils, 17–18, 21, 31, 57, 96, 125, 357–58
founder effect, 109–10
founder events, 74
FOXP2 genes, 43–45, 211
frameshift mutations, 111
Framingham (Massachusetts), 354–56
France, 37, 66, 158
Franklin, Rosalind, 16, 231n
From Hell, 189
fruit flies, 304–5
Fuegians, 129

G

G protein-coupled receptors, 86
Gacy, John Wayne, 324
Galton Collection, 232
Galton, Francis
Darwin and, 221–23, 227–28, 232–33

Galton, Francis (*continued*)
data collection by, 224–25, 264
eugenics and, 203–4, 222–23, 229, 231
Hereditary Genius, 228
on race, 219, 226–27, 249, 264
twin studies by, 297–98, 299
Galton Laboratory, 231–32
Galton, Samuel, 221
Gamow, George, 283–84
gang members, MAOA in, 316
gas chromatography, 78
gayness, 326, 327–28
gender preferences, myths about, 236–37n
genealogy testing, 50, 90, 145–47, 151–56, 167–77, 224, 305
genes
defined, 367
excerpt of, 282–83
as fate, 10, 42, 48, 168, 305, 337–38
gene chips, 296
gene flow, *20*, 50–52, 75, 94
gene therapy, 291
number of, 47, 111, 175, *270*, 271, 284
sequencing of, 7, 27–31, 41, 50, 162–63, 284, 321
See also alleles; DNA (deoxyribonucleic acid)
gene sweepstake betting book, 269–72
genetic ancestry, testing for, 50, 90, 145–47, 151–56, 167–77, 224, 305
genetic counseling, 207–8, 212, 257, 302
genetic determinism, 308, 317, 329
genetic drift, 109–10, 236, 240
genetic hitchhiking, 241–43
genetics
as astrology, 145, 166, 176–77
commercial testing for, 50, 90, 145–47, 152–56, 167–77, 224, 305
counseling for, 207–8, 212, 257, 302
defined, 367
as fate, 10, 42, 48, 168, 305, 337–38
history of, 9, 11, 48–49, 184
race and, 218–19, 234–38, 250–52
See also DNA (deoxyribonucleic acid); genes
genomes
defined, 367
human (*see* Human Genome Project)
non-human, 32
reading of, 26–30, 280–84
sequencing, 7, 27–31, 41, 50, 162–63, 284, 321
See also DNA (deoxyribonucleic acid); genes
genome-wide association studies (GWAS), 294–97, 300–301, 308, 317, 328
genomics, 6, 49, 83, 269, 273n, 348, 368
genotypes, 48, 82, 168, 234, 239, 322–23, 368–69
genus, 21–24
Germany, 23, 31, 66, 78–79, 257, 323–24, 330–31
ginger hair, 47–48, 86–93, 169, 303
Godwin's Law, 328n
gorillas, 17, 22, 59
Goss, Devon, 258
Gould, Stephen Jay, 15n
great apes (Hominidae), 22–23, 43, 59, 62
Great Britain
children per family, 353
early inhabitants, 66, 95–97, 102–4
eugenics, views on, 229–30
first cousin marriages, 206–7, 210
Galton's beauty map of, 232
genetic map of, *99*
genetic testing companies, 90, 167, 169, 171–75
infant mortality, 354
language variations from United States, 240–41, 366
milk drinking, 77–78
monarchy, 171, 185n, 194

People of the British Isles project, 97–101, 104–5, 108, 152
publicity on genetics, 88–89, 171–73, 193, 324–25
See also England
Greece, 208–209, 238
Green, Robert, 321–22
Greyfriars, 181, 183, 185
Groves, Colin, 61
Guardian, 88, 91–92, 185n, 308, 327
guns, access to, 323
GWAS (genome-wide association studies), 294–97, 300–301, 308, 317, 328
Gypsies (Roma), 119–20, 208–10

H

HACNS1, 48–49
Haeckel, Ernst, 36
hair
color of, 47–48, 82–83, 86–93, 169, 210, 298, 303
DNA extracted from, 173, 325–26
facial and head, 92–93
Handy, Benjamin, 154
HapMap, 286, 288
Happisburgh, 96
Hapsburg dynasty, 195–202, 204, 210, 303
Hapsburg Lip, 196, 303
Harris, Kelley, 53
Harry Potter books, red hair in, 87
Havasupai, 142–44, 149–50
heart disease, 110, 295, 313, 332
height, 85, 205–6, 260, 300
hemophilia, 302–3
Henry VII, King, 178, 180–81, 185n
Henry VIII, King, 180–81, 210, 303
Hepburn, Audrey, 331
HERC2, 83, 168
Hereditary Genius (Galton), 228
heritability, 296–301, 308, 328
heterozygosity, 350, 368
HEXA genes, 256
Hindus, 249
Hirschfeld, Ludwik and Hanka, 234
hitchhiking, genetic, 241–43
Hitler, Adolf, 229, 316, 325, 328n
HMS *Beagle*, 129, 220, 254
Hobbits of Flores, 24–26
Holland, 66, 174–75, 260, 313–14, 330–34
Holocaust, 230–31
Holy Prepuce, 158
Hominidae, 23, 59, 62
Homo antecessor, 24, 96n
Homo erectus, 24, 35–36, *38–39*, 66
Homo floresiensis, *20*, 24–26
Homo habilis, 23
Homo heidelbergensis, 24, 61, 66, 95
Homo neanderthalensis. See Neanderthals
Homo sapiens
emergence, 2
interbreeding with Denisovans, *20*, 60–61, 63, 261
interbreeding with Neanderthals, *20*, *38*, 51–55, 63, 73, 261
migration, 8, *38–39*
taxonomic status, 22–23
Homo sapiens sapiens, 22n
homosexuality, 326, 327–28
homozygosity, 111, 211, 368
honey, ancient use of, 80–81
honey bees, 32
Hongerwinter, 330–32, 334–35
Hood, Leroy, 271
horses, 23n, 154, 180, 260
hot sauce paradigm, 316
Hox genes, 304–5, 341
Huey, Matthew, 258
Human Genome Project
achievements of, 27, 41, 250, 273, 285, 289
challenges of, 9, 30, 32, 280–84
as mapping project, 266–68, 285
overview of, 6–7, 266–74, 367
publicity on, 285–86, 290, 307–8
hunter-gatherers, 67, 69–70, 73, 83, 136, 236–37, 245

hybrids, 23
hyoid bones, 42–43
hypervariable regions, 186
hypohidrotic ectodermal dysplasia, 243–44n

I

Ibsen, Michael, 186, 194
Iceland, 105–12, 354
ichnofossils, 96
identical by descent (IBD), 163
identical twins, 4, 87, 163, 232, 297–98, 306–7, 328
impulsivity, 316–18
inbreeding, 111–12, 199–200, 202–12, 217, 303
inbreeding coefficient (F), 200, 204
inbreeding depression, 204, 206, 210
incest, 112
Independent, 88–89, 192
India, 119–20, 153, 171, 175, 209, 249, 349–52
Indian Removal Act, 147–48
Indians, American. *See* Native Americans
infant mortality, 118, 201, 204–5, 353–56
informed consent, 143, 149–50
inhumation, 118, 136–38, 181–83, 194
intelligence, 48, 205, 272–73, 299–300, 325, 361–62
interbreeding
  by Denisovans, *20*, 60–61, 63, 261
  by *Homo sapiens*, *20*, *38*, 51–55, 60–61, 63, 73, 261
  by Neanderthals, *20*, *38*, 41, 51–55, 63, 73, 261
International Health Exhibition, 224–26
introgression, 51–54, 58–59, 103, 139–40
introns, 271n, 276, 280–81, 283–84
Inuits, 130, 131n, 139–40, 145, 236n, 248, 350–51
IQ tests, 205
Ireland, 79, 93, 108–9, 207–9
Irtysh, 68
Isabella of Portugal, 198
Ishihara test, 346
Islam, 206, 208, 251
Íslendinga-App, 112
Izzard, Eddie, 172–73

J

Jack the Ripper, misidentification of, 187–94
Jacob, François, 304
James, Oliver, 308
Japan, 148, 236n, 278, 298, 353
Jefferson, Thomas, 255
Jeffreys, Alec, 184, 192
jerks *vs.* creeps, 15
Jesus Christ, 3, 158, 287–88. *See also* Christianity
Jews, 216n, 230, 234, 256–57, 264, 287, 302, 350
Joachim Neander valley, 159
Joanna of Castille, 197, *198*
Johanson, Donald, 21
John Paul II, Pope, 59–60
Jones, Steve, 207, 336
journalism, 88–89, 166, 171–73, 192–93, 236–37, 307–8, 314, 327–28
journals, academic, 41, 166, 223, 278, 314, 316–17, 326
Judson, Horace, 277
junk DNA, 277–78
Jurassic Park, 40

K

Kalasha, 251
Kebara Cave, 42–43
KE family, 44
Kennett, Debbie, 171n
Kennewick Man, 137–39, 141, 149
kidnapping, suspicions of, 208–9
King, Turi, 181n–82n, 184–86, 193
Klinefelter's syndrome, 267, 299n

knockout/knock-in, 245
Kosminski, Aaron, 189–92
Kostenki, 67–68
Krause, Johannes, 119
Kruglyak, Leonid, 209
Kuhn, Thomas, 309
Kürten, Peter, 324

L

La Brea Tar Pits, 136, 139
lactase, 76–80
lactase persistence, 76–77, 79–80, 84
lactose intolerance, 76
Lamarckian inheritance, 336–37
Lamarck, Jean-Baptiste, 336
Landnámabók, 107
Landsker Line, 100–101
language differences, UK and US, 240–41
Lanza, Adam, 320–22
Larkin, Philip, 6
Last Glacial Maximum, 131–32n, 136
lateral gene transfer, 122
Lee, Christopher, 159–60, 161, 166
Legge, Thomas, 181
legs, genes for, 340–41
Leo III, Pope, 158
Lewontin, Richard, 235, 252
Liang Bua, 24, 26n
Lieberman, Dan, 246
life expectancy, 335, 337, 347
Linear Pottery Culture, 72, 80, 94
linkage disequilibrium, 241–42
Linnaeus, Carl, 21, 217n
Lombroso, Cesare, 329
London, 118–19, 187–90, 217, 256, 347
Longlin Cave, 61
Loschbour, 72–73, 83, 126, 140
loss of function mutations (LOFs), 111
Louhelainen, Jari, 189–92
Lucy, 21–22
Lunar Society, 221
Luxembourg, 72, 83, 126
Lykken, David, 299n

M

macroevolution *vs.* microevolution, 359
macular degeneration, 295–96
magnetic resonance imaging (MRI), 323
Mail on Sunday, 189–91, 193
malaria, 237–38, 340, 350
malnutrition, effects of, 330–32, 334–35
Malthus, Thomas, 230
mammals, 75, 123, 275, 334, 341–42, 345
Manhattan plots, 294, 296, 317
Manica, Andrea, 125n–26n
mantis shrimp, 345, 361
manual dexterity, 49, 273
MAOA (monoamine oxidase A), 312–17
Maori, 142, 314–15
Margarita of Austria, 197, *198*
Maria Anna of Neuburg, 196
Maria Anna of Spain, 197, *198*
Mariana of Austria, 196, 197, *198*
Marie Louise of Orléans, 196
Markow, Therese, 143
marmots, 113–14
marriages, first cousin, 199–200, 203–4, 206–7, 210–11
mass shootings, 320–23
maternal malnutrition, 331–32, 334
Mathieson, Iain, 83–84
MC1R genes, 47–48, 86–88, 91n, 92–93, 369
McCann, Madeleine, 208
McQuillan, Ruth, 206
media coverage, 88–89, 166, 171–73, 192–93, 236–37, 307–8, 314, 327–28
meiosis, 6
melanin, 82–83, 86–87
melanocortin 1 receptor (MC1R), 47–48, 86–88, 91n, 92–93, 369
melanocytes, 82, 86, 88
Mencken, H. L., 265, 326

Mendel, Gregor, 16, 202, 233, 304, 306–7, 328, 368
Mendel's Laws, 233, 292, 304, 306, 368
men, of eminence, 228
mental illness, 195–96, 197–98, 230–31, 318–19, 321
methyl tags, 332–33, 337–38
mice, 45, 202, 245, 273, 333, 336–37
microevolution *vs.* macroevolution, 359
microphthalmia, 211, 275
microsatellites, 288–89
Middle East, 2, 70, 79, 83, 211, 238, 251
migration, *38–39*, 49–50, 73–75, 124, 134, 140–43, 242, 260–61
milk, human consumption of, 75–81, 237
minisatellites, 289
mitochondria, 5, 34–35, 186, 368–69
mitochondrial DNA (mtDNA)
 defined, 368–69
 extracted from dirt, 64n
 from Iceland, 107–9
 of Jack the Ripper (purported), 191–92
 Native American, 146
 Neanderthal, 37–39, 40, 53
 of Prince William, 171
 of Richard III, 185–86, 192
 transmission of, 34–35, 145
Moffat, Alistair, 89–90, 171
monarchy, British, 171, 185n, 194
Mongolia, 236n, 249
monkeys, 23, 314, 345. *See also* apes; chimpanzees
monoamine oxidase A (MAOA), 312–17
monozygotic twins, 297
Moore, Alan, 189n
Morgan, Thomas Hunt, 304
mosquitoes, 238, 348
most recent common ancestor (MRCA), 122, 161, 164
Mota cave, 125n
Motala, 72–73, 83–84, 126
mtDNA. *See* mitochondrial DNA (mtDNA)
Mulligan, Connie, 141
multiregional hypothesis, 261
Muslims, 205, 206–7, 249. *See also* Islam
mutations
 314.1C and 315.1C, 192
 370A, 244–47
 color blindness, 345–46
 FOXP2, 43–45, 211
 frameshift, 111
 HEXA, 256
 impact on proteins, 44n, 110–11, 244
 lactase, 77, 79
 loss of function (LOFs), 111
 malaria, 237–38
 random, 332
 red hair, 86–88, 91–93
 role in evolution, 44n, 347–48
Myers, P.Z., 322
MYL4 genes, 110

N

Naming Jack the Ripper (Edwards), 189
Napoleon, 173, 325
Native American DNA (TallBear), 141n
Native American Graves Protection and Repatriation Act (NAGPRA), 137–39
Native Americans
 animosity towards science, 142–44, 148–50
 genetic testing for, 145–47
 genomes of, 135, 138–42, 146–47, 155
 mitochondrial DNA, 146
 racism and prejudice against, 147–50, 230, 255
 remains of, 136–39
 as term, disagreement on, 131
 tribal status, 144–45, 147–48
 See also Americas

natural selection, 52–53, 84–86, 89–90, 123–24, 219n–20, 336–37, 352–60. *See also* evolution; mutations
Nature (journal), 24–25, 41, 56, 223, 231n, 278, 286, 290n, 316
Nature Communications, 193
nature *vs.* nurture, 223, 297, 299, 333
Nature via Nurture (Ridley), 299
Nazis, 230–31, 257, 325, 330–31, 334
Neanderthals
 author's genome and, 170, 175
 brain capacity of, 36
 culture of, 37
 fossil remains, 8, 17, 31–33, 35–36, 51–52, 95
 genome, 42–45, 47–50, 51–54, 88, 170
 interbreeding by, *20*, *38*, 41, 51–55, 63, 73, 261
 migration of, *38–39*, 66
 mitochondrial DNA, 37–40, 53
 physical appearance, 36, 50
 speech, capacity for, 42–45
 taxonomic status, 23
 type specimen, 31–32
negroes, prejudiced views on, 223n, 226, 234n, 249, 257, 263
Netea, Mihai, 120
Netherlands, 66, 174–75, 260, 313–14, 330–34
Neumann, Joachim, 159
neurotransmitters, 312–13
Neville, Cecily, 179, 185
New York Times, 144, 149, 247, 321–22
New Zealand, 142, 216–17, 314–15
Nielsen, Rasmus, 53, 140
Nijmegen, 313
non-coding DNA, 277–79
North America, 74, 106, 133, 136, 141, 151–52
Norway, 100, 103–4, 108, 354
noses, 45–47, 237, 276, 343. *See also* smell, sense of
nucleotides, 27, 29, 366, 367, 369
Nüsslein-Volhard, Christiane, 121n
Nuu-chah-nulth, 143–44

O

Obama, Barack, 139
obesity, 350–51
O'Briain, Dara, 309
O'Byrne, Kevin, 153–54
oculocutaneous albinism, 210
odds ratios, 294n
Ohno, Susumu, 278
olfaction. *See* smell, sense of
Olivier, Laurence, 179, 181
Olympics, black people in, 257
On the Origin of Species (Darwin), 1, 3–4, 16, 220, 254, 263
Operation Manna and Chowhound, 331
opsins, 344–46
OR7D4 genes, 46–47
orangutans, 17, 59, 95
Orcadians, 100, 108
orchids, 204
Oriental rat fleas (*Xenopsylla cheopsis*), 113–14, 117
Otto, John, 303
Out of Africa hypothesis, 36, *38*, 49, 52, 124, 260–61
Överkalix, 335, 337
Owens, Jesse, 257

P

Pääbo, Svante, 32, 37, 40, 42, 56, 64n, 68
Pakistanis, in Britain, 206–7, 215, 217
paleoanthropology, 32, 36, 42, 56
Paleo Diet, 69–70
panglossianism, 237, 343
paternity, false, 185
Pax genes, 304–5
pedigree collapse, 199–200
pedigrees, 2, 159–61, 197, 199–200
penetrance, 328
People of the British Isles project, 97–101, 104–5, 108, 152

petrous bone, 84
phaeomelanin, 82, 87
phenotypes, 48, 82, 93, 239, 306, 318, 369
Philip I (the Handsome), 197, *198*, 200
Philip II of Spain, *198*, 201
Philip III of Spain, *198*, 200
Philip IV of Spain, 197, *198*
photoreceptors, 344–45
phrenology, 329–30
phyletic gradualism, 15n
pigeons, 222, 227–28
pigs, smell of, 46–47
Piltdown Man, 95
pink, gender preferences for, 236–37n
Pippin the Short, 158
plague, 115–24, 157
plant genomics, 273n
Plasmodium, 237
pneumonic plague, 115, 123
polymorphisms, 288, 328, 369
prejudice. *See* racism and prejudice
primeval soup, 277
Princes in the Tower, 179–80
Private Eye, 268
Procopius, 116–17, 121
PROP1 genes, 201
proteins, 16, 86–88, 110–11, 271–77, 280–81, 367, 369
pseudogenes, 276
punctuated equilibrium, 15n
Punnett squares, 306

Q

"quantum," as buzzword, 334
Queen of Sheba, 171–72

R

race
  categorizations of, 249–52
  concept of, 9, 248–250, 262
  Darwin's views on, 214, 223, 262–63
  disease and, 237–38, 256–57
  Galton's views on, 219, 226–27, 249, 264
  genetics and, 218–19, 234–38, 250–52
  nonexistence of, 218–19, 250, 261–62
  sports and, 257–60
  Wade's views on, 247–48, 250
  See also black people
racism and prejudice
  author's experience of, 214–15
  of Francis Galton, 219, 226–27, 249, 264
  against Maori, 315
  against Native Americans, 147–50, 230, 255
  against Roma, 208–9
  slavery as, 130–31, 152, 223n, 227, 259–60
Ralph, Peter, 163
Rasmussen, Simon, 121–22
rats, 117, 124, 333, 334–35
recessive alleles, 200, 305–7, 368
recessive diseases, 111, 200–202, 206–8, 210–12, 256, 291
recombination, 35, 52–53
Red Deer Cave people, 61
red hair, 47–48, 86–93, 169, 303
Reich, David, 60–61, 68, 72–73
religion and science, bridging of, 59–60
repeats, 276–77
replication, DNA, 347
ribonucleic acid (RNA), 271, 276–77, 290, 366, 367
rice, 273
Richard III, King, 7, 178–87, 192–93, 194–95
Richard Tertius, 181
Richmond (Henry VII), 178, 180–81, 185n
Ridley, Matt, 299n
Ring a Ring of Roses, 115n
risk-taking, 317
RNA (ribonucleic acid), 271, 276–77, 290, 366, 367

Roberts, Richard, 271n
rods, retinal, 275, 344–45
Roma (Gypsies), 119–20, 208–10
Roman Empire, 71, 101, 116, 157, 174
Romania, 78, 120–21
Romans, 78, 96–97, 101–4
Roosevelt, Theodore, 229
Rosenberg, Noah, 250, 252
Rowen, Lee, 270–72, 284
Rowling, J. K., 87
Rumsfeld, Donald, 272
Russia, 55–56, 67–68, 131–32, 355
Rutherford, Adam (author), 92–93, 106, 168–170, 214–18, 253, 328n
Rutherford, Adam (Egyptologist), 106n
Rutherford family (author), 92, 153–54, 169, 215–18
Rutherford's Law, 328n
ruthlessness, genes for, 316

S

Sachs, Bernard, 256
Sagan, Carl, 133n, 253
salivary amylase, 69
Samaritans, 164
sampling errors, 218
Sanders, Edward, 63
Sandy Hook shootings, 320
Sanger, Fred, 27, 30
Saxons, 96, 101–3, 141, 158
Scandinavia, 79, 91, 94, 103, 106, 108–9, 170, 353–54. *See also* Denmark; Norway; Sweden
schizophrenia, 143, 231, 332
Schürer, Kevin, 182n, 185n
Science (journal), 41, 125n, 290n, 314
science and religion, bridging of, 59–60
science books metaphor, 253–55
scoliosis, 183
Scotland, 70, 82, 89–92, 101, 108
ScotlandsDNA, 89–91, 173
Secret History (Procopius), 116–17, 121
selective sweep, 120, 140, 241–42
sensory processing disorder, 321
septicemic plague, 115, 123
sequencing, 7, 27–31, 41, 50, 162–63, 284, 321
sex (gender), 33, 57, 299
sex (reproduction), *20*, 49–51, *62*–63, 152, 237
sex chromosomes, 33–35, 183, 304, 366–67. *See also* X chromosomes; Y chromosomes
sexual abuse, 315
sexuality, 326, 327–28
Shakespeare, William, 178–82, 187, 194
Shaw, George Bernard, 230, 240
Sheba, Queen of, 171–72
shootings, mass, 320–23
shotgun sequencing, 284
shoveling/scalloping, 244, 362
Siberia, 51, 55, 57, 60, 74–75, 132–33, 144
sickle cell anemia, 237–38, 348, 350
single nucleotide polymorphisms (SNPs), 84, 120, 167–68, 240–41, 244, 348, 369
skin color, 74, 81–85, 89, 208–10, 248–50, 257–60, 264
Skraeling, 128–29, 130
Slaughterhouse-Five (Vonnegut), 14, 298n
slavery, 130–31, 152, 223n, 227, 259–60
SLC24A5 and SLC45A2, 83
smallpox, 146, 303, 340
smell, sense of, 45–47, 237, 242–43, 276, 343
SNPs (single nucleotide polymorphisms), 84, 120, 167–68, 240–41, 244, 348, 369
Somerset, Henry, 185
South America, 130, 134–35, 138, 164–65
South Asians, 130, 212
South Korea, 240, 243, 248
South Park, 91n

Spain, 37, 83–85, 88, 164, 195–202, 210
species, classification of, 21–24
speech, capacity for, 42–45, 211
sperm, 4–6, 34, 52, 175, 275, 299
Spider-Man!, 342n–43n
splicing, alternative, 290
sports, 257–60
SRGAP2, 48
Standard Model, 303–4
starvation, effects of, 330–32, 334–35, 337
statistics, 218, 219n, 233, 294n, 317–23
Stefánsson, Kári, 108, 110
sterilization, forced, 148, 229–31
Stopes, Marie, 229
stress, effects of, 333, 334–35
Stringer, Chris, 61, 96n
STRUCTURE program, 251
Sturtevant, Alfred, 307
Stuttgart, 72–73, 140
subspecies, 22n
succinylcholine, 349
Sulston, John, 271n, 289
superheroes, 339–40, 342n–43n
surfing, genetic, 242–43
Sutikna, Thomas, 26n
sweat glands, 243–45, 247
Sweden, 72, 82, 83, 126, 230, 317, 335, 353–55
Szathmáry, Em ke, 141

T

Taíno, 129–30
TallBear, Kim, 141n
Talmud, genetic advice in, 302
taxonomy, 21–24
Tay, Waren, 256
Tay-Sachs disease, 211, 256–57, 264
TBX15 genes, 139
teeth
  DNA extracted from, 25, 72–73, 79, 108, 117, 121, 146
  shoveling/scalloping in, 244, 362
testing companies. *See* 23andMe; Accu-Metrics; BritainsDNA
tetrachromacy, 346
theory, defined, 359–60
Thingvellir, Baron, 105–6, 108–9
Thomas, Mark, 79–80, 177, 182n
Tibet, 58–59, 258, 264
Tiihonen, Jari, 317–18
Tilousi, Edmond, 144
Times, 171, 222n, 226, 247
Toll-like receptors (TLRs), 120–21
tongue rolling, 306–7
transcription factors, 139, 275–76
transgenerational epigenetics, 335–38
transitional fossils, 357–58
tribal status, 144–45, 147–48
A Troublesome Inheritance (Wade), 247–48
Tsui, Kyle, 147
Tudors, 179–81, 185n, 194
Turner, John, 15n
Turner's syndrome, 267, 299n
Twain, Mark, 226
twin studies, 87, 93, 232, 297–98, 299–300, 306–7, 328
type specimens, 31–32

U

uniqueness, 15, 362
United Kingdom. *See* Great Britain
United States
  African Americans, 131, 152, 214, 217, 257–58
  eugenics, history of, 148, 230
  first cousin marriages, 210
  Framingham study, 354–55
  genetic testing companies, 145–47
  genomes of, 151–52
  indigenous population, 130–36, 139–40, 142–44, 146, 149–50
  infant mortality, 354
  language variations from Great Britain, 240–41, 366
  Native Americans (*see* Native Americans)
  white Americans, 255
  See also Americas

University College London, 44, 79, 223, 231
Ust'-Ishim, 68

V

Vaishyas, 349–50
Val60Leu, 92–93
Vancini, Rodrigo, 259
Venter, Craig, 285, 289n–90n
verbal dyspraxia, 44
Vespucci, Amerigo, 129
video games, violence in, 320–21
Vikings, 100, 103–6, 108, 127, 128–29, 170, 172
violence, 188–89, 313–21, 324
virgin birth, 287
vitamin D, 79, 84, 89, 236
Voltaire, 237, 343
Vonnegut, Kurt, 14, 298n

W

Wade, Nicholas, 247–48, 250
Waldroup, Davis Bradley, 311–12, 319
Waldroup, Penny, 311–12
Wales, 35, 66, 95, 100–101, 105
Walsh, Anthony, 317
Ward, Ryk, 144
warrior gene, 314–17, 319
Watson, James, 16, 231n, 233, 269, 274, 283, 289
Wedgwood family, 204–5, 227
Weisman, August, 336
Wellcome Trust, 30n, 289, 295
Whitener, Ron, 149
Wilde, Oscar, 240n
William, Prince, 171
Wilson, Jim, 172
wings, 341–42
Woolas, Phil, 207
World War II, 148, 230–31, 330–31, 334
writing, as start of history, 2

X

X chromosomes, 33–34, 54, 299n, 302, 313, 333, 345–46, 367
Xenopsylla cheopsis, 113–14, 117
XXY syndrome, 267, 299n
XYY syndrome, 267

Y

yam farming, 238
Yamnaya, 74, 84, 85
Y chromosomes, 5, 33–35, 54, 57–58, 107–8, 173–75, 183–85, 367
Yersinia murine toxin (Ymt), 122
Yersinia pestis, 113–17, 119, 121–24
Yersinia pseudotuberculosis, 122

Z

zebra finches, 45
zombie arguments, 357

ABOUT THE AUTHOR

Adam Rutherford is a science writer and broadcaster. He studied genetics at University College London, and during his PhD on the developing eye, he was part of a team that identified the first genetic cause of a form of childhood blindness. He has written and presented many award-winning series and programs for the BBC, including the flagship weekly Radio 4 program *Inside Science*, *The Cell* for BBC Four, and *Playing God* (on the rise of synthetic biology) for the leading science series *Horizon*, as well as writing for the science pages of the *Guardian*. His first book, *Creation*, on the origin of life and synthetic biology, was published in 2013 to outstanding reviews and was short-listed for the Wellcome Trust Prize.